Holt Algebra 2

Chapter 13 Resource Book

HOLT, RINEHART AND WINSTON

A *Harcourt* Education Company

Orlando • Austin • New York • San Diego • London

ISBN 978-0-03-042823-4
ISBN 0-03-042823-8

6 7 8 9 862 09

Contents

Holt Algebra 2

Date ______________

Dear Family,

In Chapter 13 your child will begin to learn about trigonometry. The word *trigonometry* comes from Greek words meaning "triangle measurement."

A **trigonometric function** is defined by a rule that compares the lengths of two sides of a right triangle. The fundamental trigonometric functions are **sine, cosine,** and **tangent**. Three additional functions—**cosecant, secant**, and **cotangent** are defined as reciprocals. In trigonometry, θ, is traditionally used to represent an angle measure.

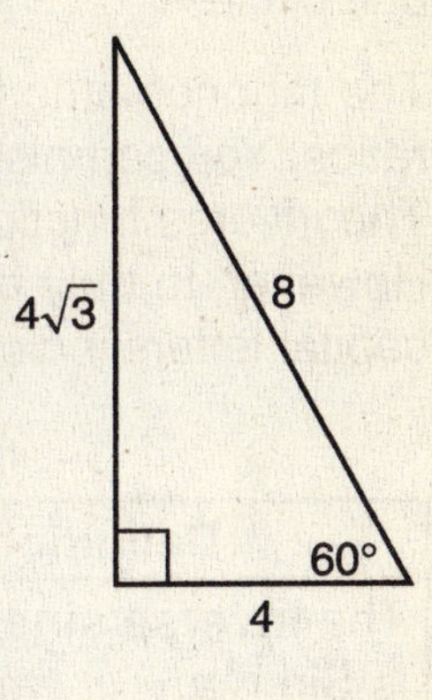

Trigonometric Functions		
Definition	**Symbols**	**Example**
The **sine** of angle θ is the ratio of the length of the opposite leg to the length of the hypotenuse.	$\sin\theta = \dfrac{\text{opp.}}{\text{hyp.}}$	$\sin 60° = \dfrac{4\sqrt{3}}{8} = \dfrac{\sqrt{3}}{2}$
The **cosine** of angle θ is the ratio of the length of the adjacent leg to the length of the hypotenuse.	$\cos\theta = \dfrac{\text{adj.}}{\text{hyp.}}$	$\cos 60° = \dfrac{4}{8} = \dfrac{1}{2}$
The **tangent** of angle θ is the ratio of the length of the opposite leg to the length of the adjacent leg.	$\tan\theta = \dfrac{\text{opp.}}{\text{adj}}$	$\tan 60° = \dfrac{4\sqrt{3}}{4} = \sqrt{3}$
The **cosecant** of angle θ is the reciprocal of the sine function.	$\csc\theta = \dfrac{1}{\sin\theta} = \dfrac{\text{hyp.}}{\text{opp.}}$	$\csc 60° = \dfrac{8}{4\sqrt{3}} = \dfrac{2}{\sqrt{3}} = \dfrac{2\sqrt{3}}{3}$
The **secant** of angle θ is the reciprocal of the cosine function.	$\sec\theta = \dfrac{1}{\cos\theta} = \dfrac{\text{hyp.}}{\text{adj.}}$	$\sec 60° = \dfrac{8}{4} = 2$
The **cotangent** of angle θ is the reciprocal of the tangent function.	$\cot\theta = \dfrac{1}{\tan\theta} = \dfrac{\text{adj.}}{\text{opp.}}$	$\cot 60° = \dfrac{4}{4\sqrt{3}} = \dfrac{1}{\sqrt{3}} = \dfrac{\sqrt{3}}{3}$

In a right triangle, the non-right angles must be acute, or $0° < \theta < 90°$. However, trigonometric functions can be evaluated for any real value of θ. Any positive or negative angle measure can be thought of as an **angle of rotation**, with the **initial side** on the positive *x*-axis and the **terminal side** rotated either counterclockwise (positive) or clockwise (negative). Angles beyond 360° and −360° can also be created with multiple revolutions.

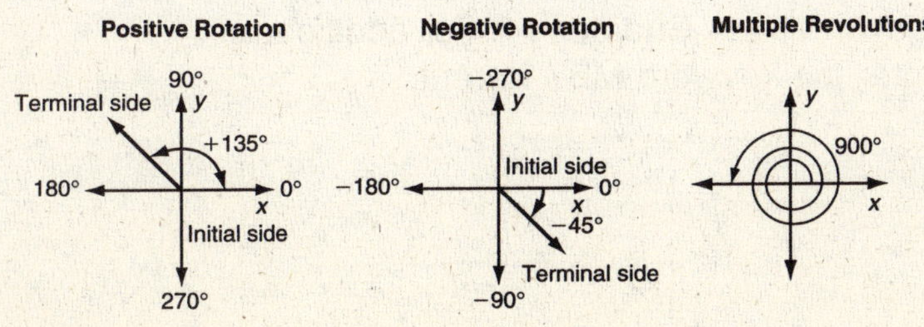

Holt Algebra 2

Most people are familiar with measuring angles with degrees. Another unit of angle measure is the **radian**, which is based on arc length. A complete circle has circumference $2\pi r$ and 360°, so radians and degrees are related by the ratio $\dfrac{2\pi \text{ radians}}{360°}$, or $\dfrac{\pi \text{ radians}}{180°}$.

A **unit circle** is a circle centered at the origin of a coordinate grid with radius 1 unit. The unit circle at right shows several special angles measured in degrees and radians, and the corresponding x- and y-coordinates of points on the unit circle. You may notice that there is a lot of symmetry between the quadrants, which will help your child memorize the unit circle.

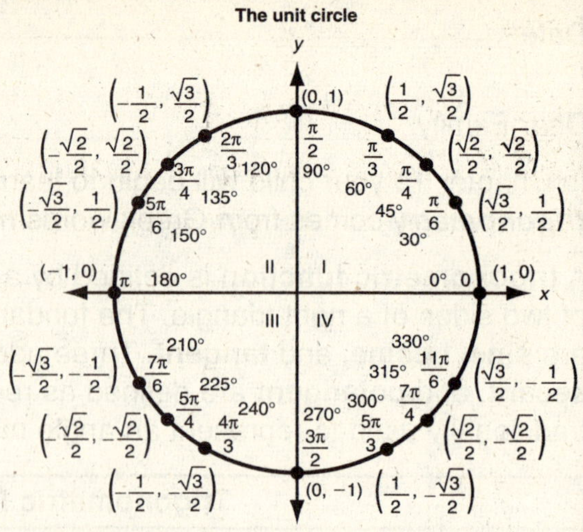

The trigonometric functions use angle measures as inputs and output ratios. You can also use inverse trigonometric functions to work in reverse. The inverse functions use ratios as inputs and output angle measures. However, to truly be functions, the inverses have restricted domains. (The capital letters indicate that the functions have restricted domains.)

Inverse Trigonometric Functions

Definition	Domain	Range
The **inverse sine** function is $\text{Sin}^{-1}a = \theta$, where $\sin\theta = a$.	$\{a \mid -1 \le a \le 1\}$	$\left\{\theta \mid -\frac{\pi}{2} \le \theta \le \frac{\pi}{2}\right\}$ $\{\theta \mid -90° \le \theta \le 90°\}$
The **inverse cosine** function is $\text{Cos}^{-1}a = \theta$, where $\cos\theta = a$.	$\{a \mid -1 \le a \le 1\}$	$\{\theta \mid 0 \le \theta \le \pi\}$ $\{\theta \mid 0° \le \theta \le 180°\}$
The **inverse tangent** function is $\text{Tan}^{-1}a = \theta$, where $\tan\theta = a$.	$\{a \mid -\infty < a < \infty\}$	$\left\{\theta \mid -\frac{\pi}{2} < \theta < \frac{\pi}{2}\right\}$ $\{\theta \mid -90° < \theta < 90°\}$

The chapter concludes with two important laws that can be used to "solve" a triangle, which means to find all of the angle measures and side lengths when only a few are known.

Law of Sines: $\dfrac{\sin A}{a} = \dfrac{\sin B}{b} = \dfrac{\sin C}{c}$

Law of Cosines: $a^2 = b^2 + c^2 - 2bc\cos A$

$b^2 = a^2 + c^2 - 2ac\cos B$

$c^2 = a^2 + b^2 - 2ab\cos C$

You child will continue to study trigonometry in Chapter 14. For additional resources, visit go.hrw.com and enter the keyword MB7 Parent.

Holt Algebra 2

Practice A
Right-Angle Trigonometry

Find the value of the sine, cosine, and tangent functions for θ.

1.

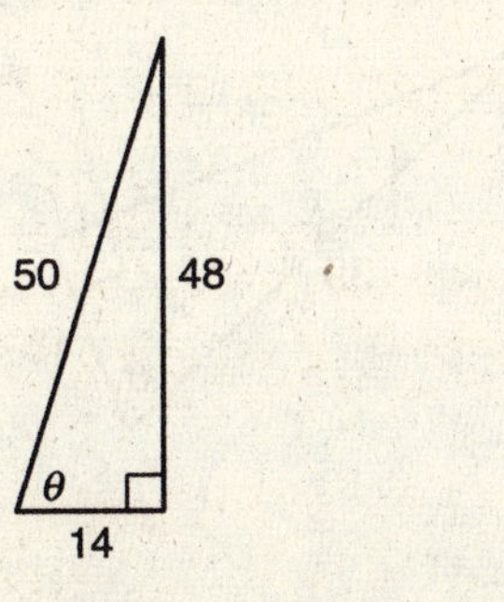

Write and simplify the fractions.

a. $\sin \theta = \dfrac{\text{opp.}}{\text{hyp.}} =$ _______________________

b. $\cos \theta = \dfrac{\text{adj.}}{\text{hyp.}} =$ _______________________

c. $\tan \theta = \dfrac{\text{opp.}}{\text{adj.}} =$ _______________________

2.

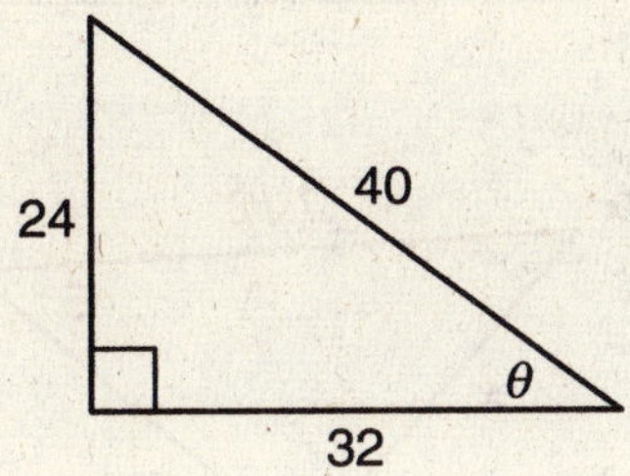

3.

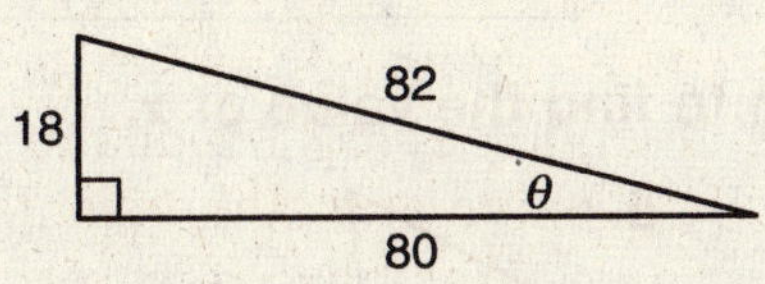

4.

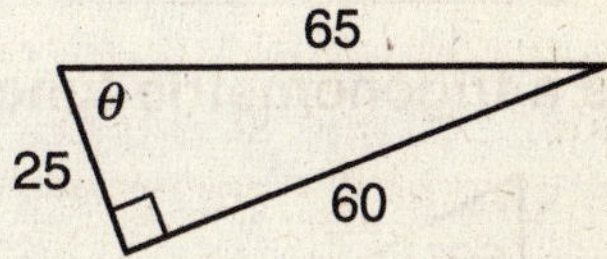

_______________   _______________   _______________

Use a trigonometric function to find the value of x.

5.

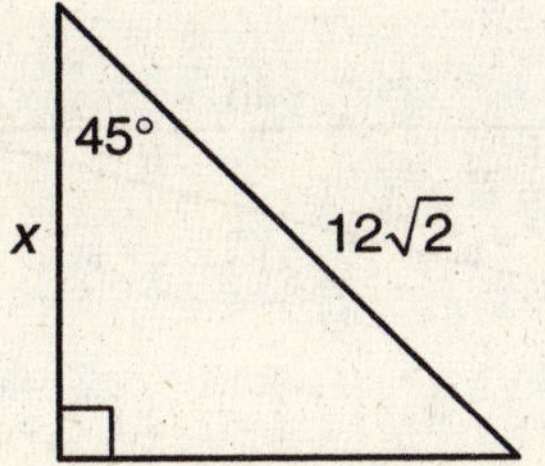

a. Choose the function.   _______________

b. Substitute given values.   _______________

c. Evaluate the trigonometric
ratio for the angle measure.   _______________

d. Solve the proportion for x.   _______________

6.

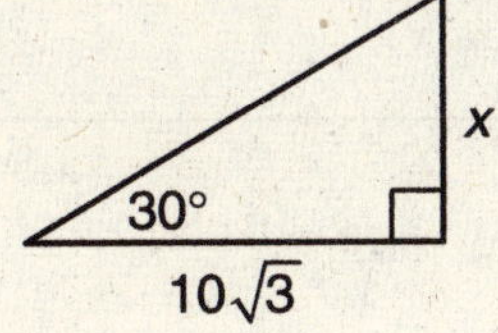

7.

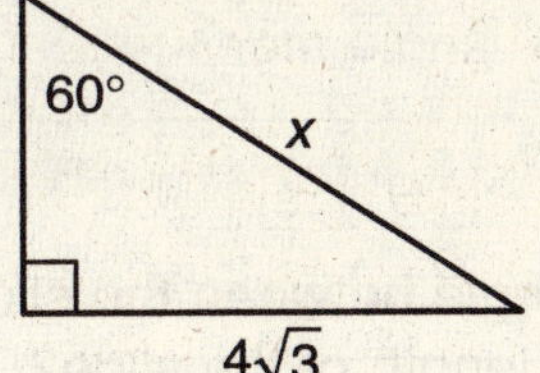

8.

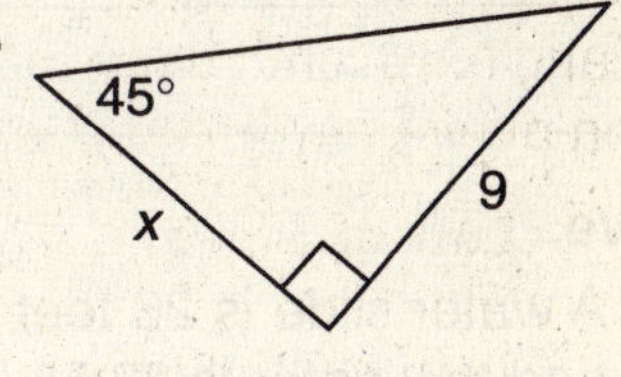

_______________   _______________   _______________

Solve.

9. A conveyor belt leads from the ground to a barn door 24 feet high.
The angle between the belt and the ground is 32°. What is the
length of the conveyor belt to the nearest foot?

____

Holt Algebra 2

Practice B
LESSON 13-1 *Right-Angle Trigonometry*

Find the value of the sine, cosine, and tangent functions for θ.

1.

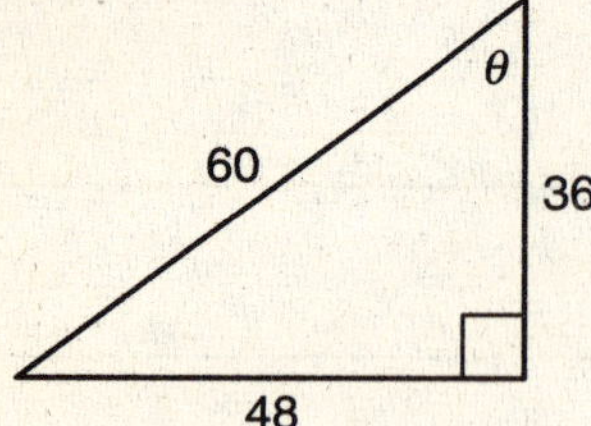

2.

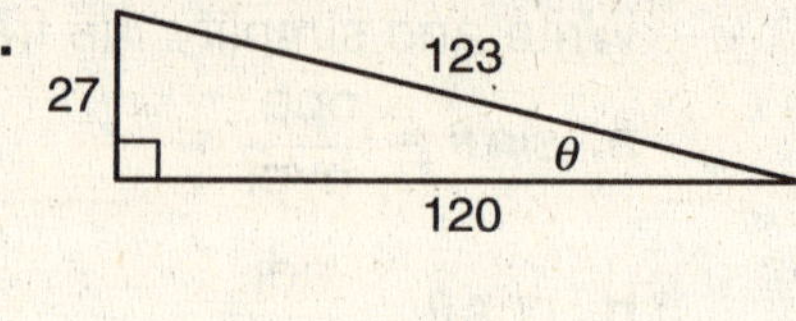

3.

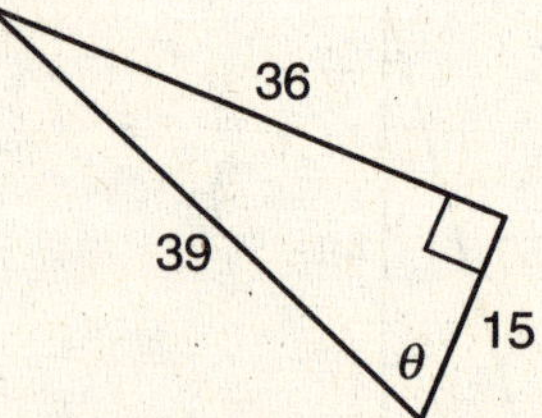

_______________   _______________   _______________

Use a trigonometric function to find the value of x.

4.

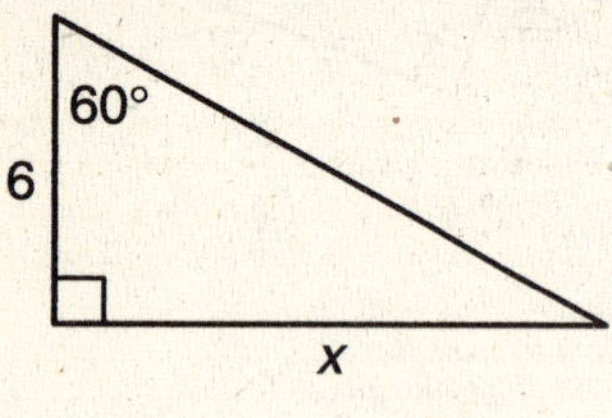

5.

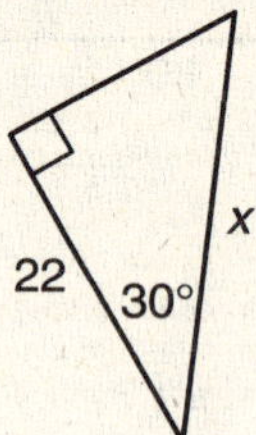

6.

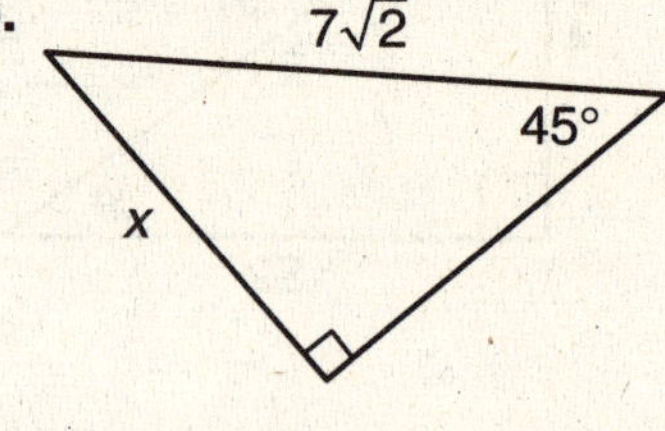

_______________   _______________   _______________

Find the values of the six trigonometric functions for θ.

7.

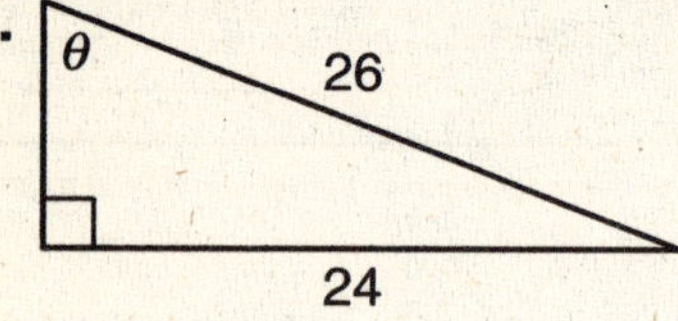

8.

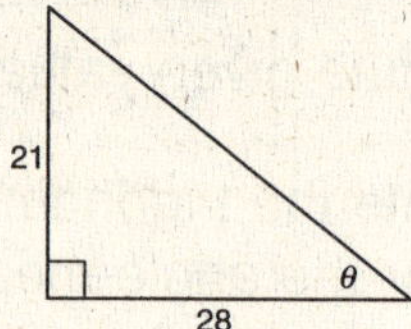

9.

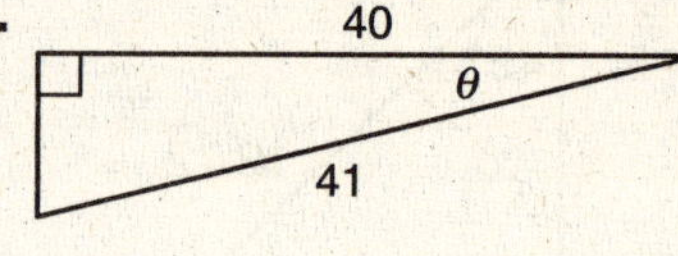

_______________   _______________   _______________

_______________   _______________   _______________

Solve.

10. A water slide is 26 feet high. The angle between the slide
and the water is 33.5°. What is the length of the slide?   _______________

11. A surveyor stands 150 feet from the base of a viaduct and
measures the angle of elevation to be 46.2°. His eye level
is 6 feet above the ground. What is the height of the viaduct
to the nearest foot?   _______________

12. The pilot of a helicopter measures the angle of depression to
a landing spot to be 18.8°. If the pilot's altitude is 1640 meters,
what is the horizontal distance to the landing spot to the
nearest meter?   _______________

Holt Algebra 2

LESSON 13-1 — **Practice C**
Right-Angle Trigonometry

Find the value of the sine, cosine, and tangent functions for θ.

1.
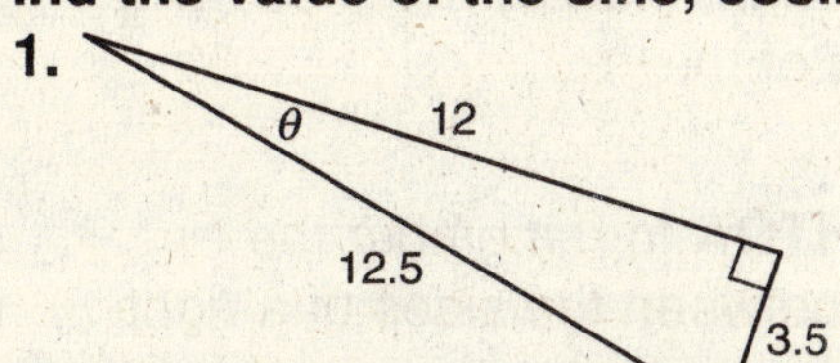

2.
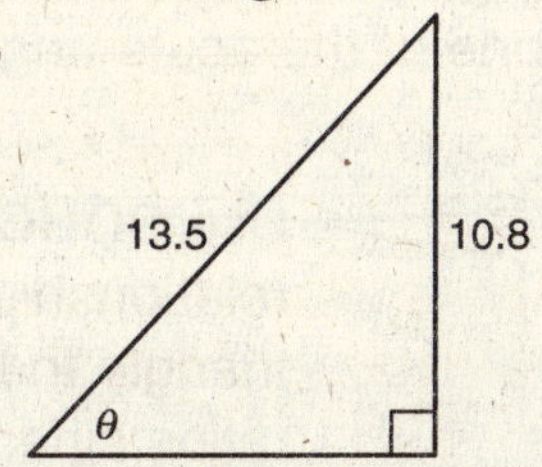

3.
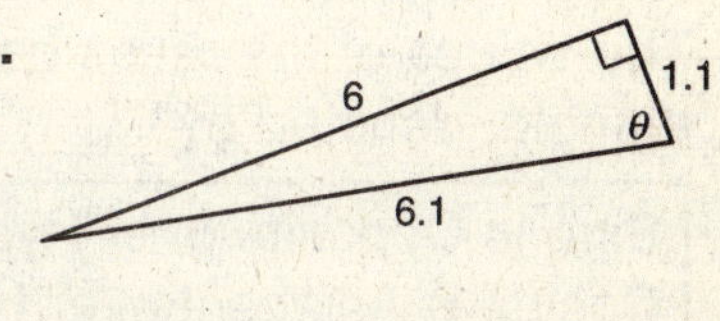

_______________________   _______________________   _______________________

Use a trigonometric function to find the value of *x*.

4.
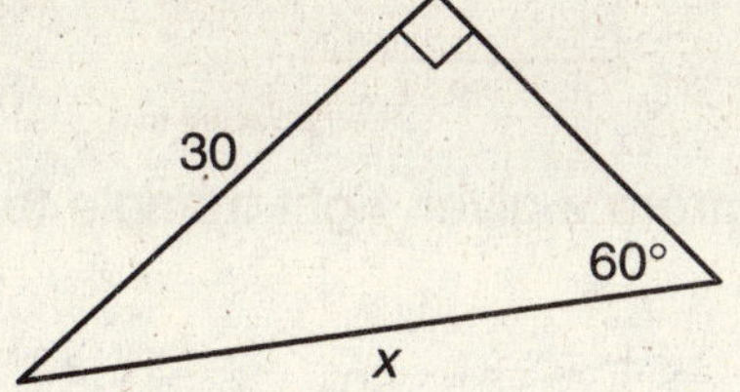

5.
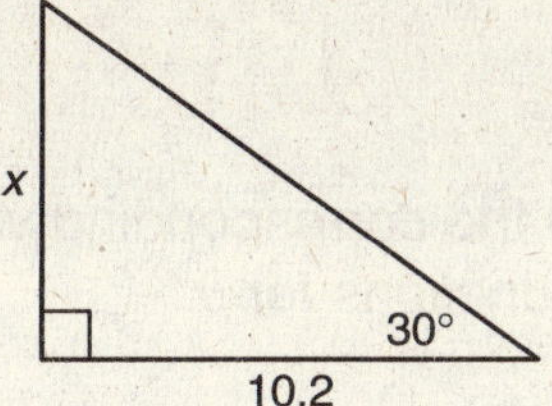

6.
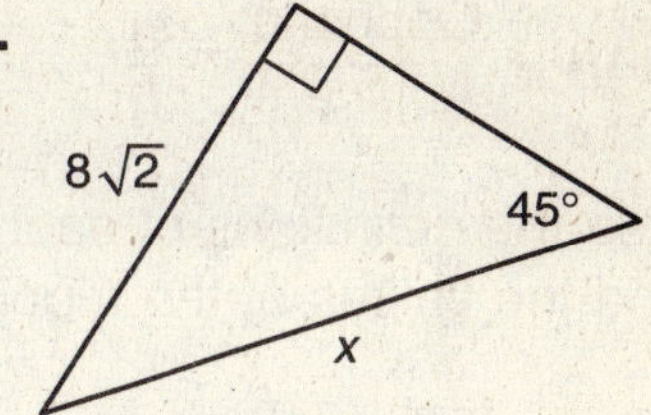

_______________________   _______________________   _______________________

Find the values of the six trigonometric functions for θ.

7.
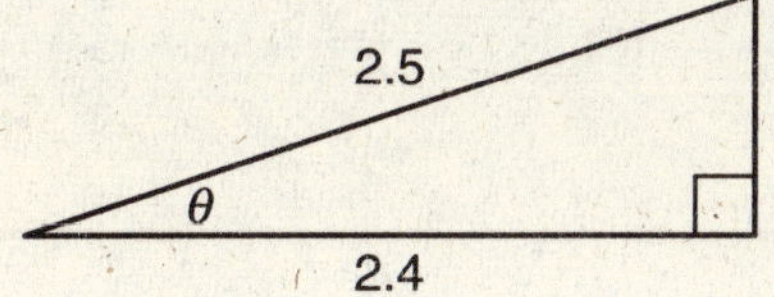

8.
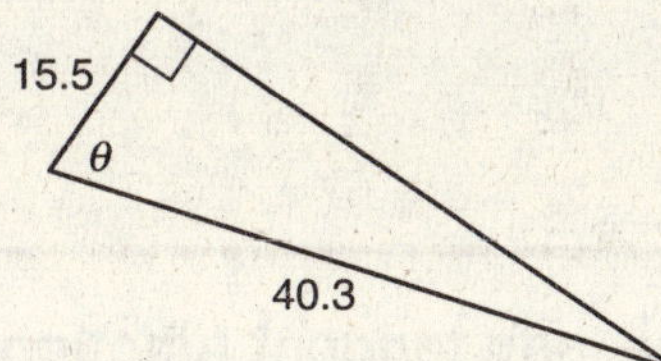

9.
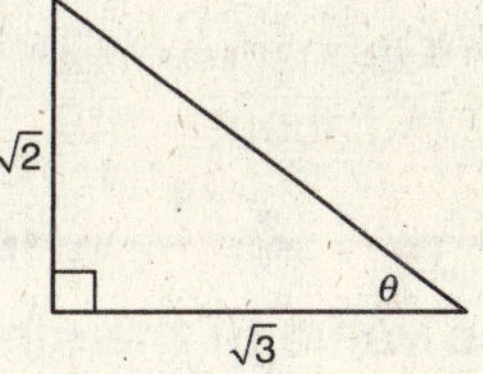

_______________________   _______________________   _______________________

_______________________   _______________________   _______________________

Solve.

10. A kite string is 102 feet long. The angle between the
kite string and the ground is 54.9°. How high is the kite?   _______________________

11. A surveyor stands 186 feet from the base of a cliff and
measures the angle of elevation to be 56.6°. His eye level
is 5 feet above the ground. What is the height of the cliff
to the nearest foot?   _______________________

12. The pilot of a hot air balloon measures the angle of
depression to a landing spot to be 36.7°. If the pilot's
altitude is 1752 meters, what is the horizontal distance
to the landing spot to the nearest meter?   _______________________

Holt Algebra 2

Reteach
Right-Angle Trigonometry

A **trigonometric ratio** compares the lengths of two sides of a right triangle.
The values of the ratios depend upon one of the acute angles of the
triangle, denoted by θ.

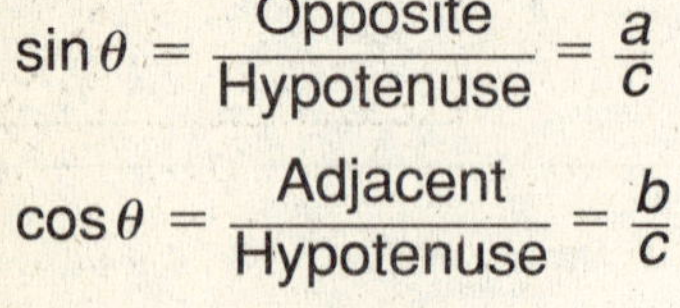

Use **SOHCAHTOA** to remember the
relationships between the sides of a right
triangle that correspond to the trigonometric
ratios sine, cosine, and tangent.

$$\sin\theta = \frac{\text{Opposite}}{\text{Hypotenuse}} = \frac{a}{c}$$

$$\cos\theta = \frac{\text{Adjacent}}{\text{Hypotenuse}} = \frac{b}{c}$$

$$\tan\theta = \frac{\text{Opposite}}{\text{Adjacent}} = \frac{a}{b}$$

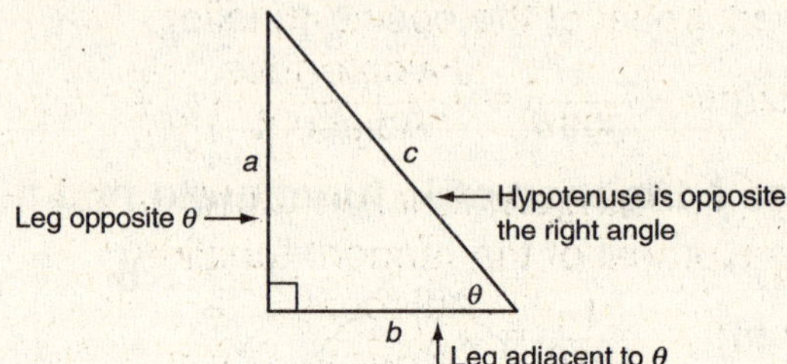

Use the definitions of each ratio and the corresponding values from a given right triangle to
find the values of the trigonometric functions for θ.

$$\sin\theta = \frac{\text{Opposite}}{\text{Hypotenuse}} = \frac{14}{50} = \frac{7}{25}$$

$$\cos\theta = \frac{\text{Adjacent}}{\text{Hypotenuse}} = \frac{48}{50} = \frac{24}{25}$$

$$\tan\theta = \frac{\text{Opposite}}{\text{Adjacent}} = \frac{14}{48} = \frac{7}{24}$$

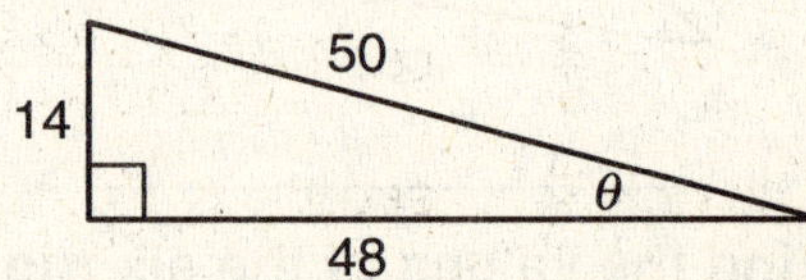

Find the value of the sine, cosine, and tangent functions for θ.

1.
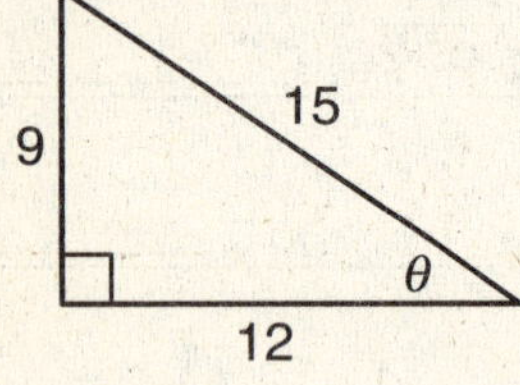

2.
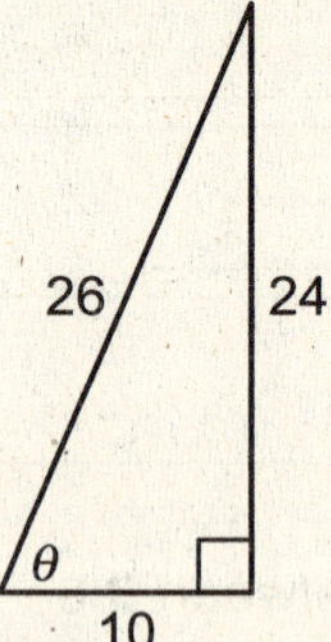

3.
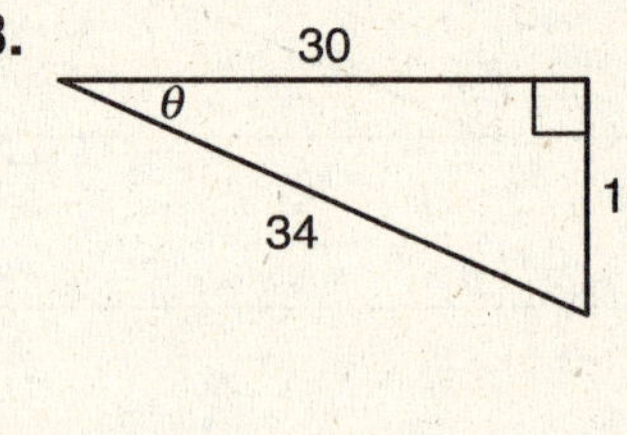

$$\sin\theta = \frac{\text{Opposite}}{\text{Hypotenuse}} = \underline{\hspace{2cm}}$$

$$\cos\theta = \frac{\text{Adjacent}}{\text{Hypotenuse}} = \underline{\hspace{2cm}}$$

$$\tan\theta = \frac{\text{Opposite}}{\text{Adjacent}} = \underline{\hspace{2cm}}$$

$\sin\theta = \underline{\hspace{2cm}}$

$\cos\theta = \underline{\hspace{2cm}}$

$\tan\theta = \underline{\hspace{2cm}}$

$\sin\theta = \underline{\hspace{2cm}}$

$\cos\theta = \underline{\hspace{2cm}}$

$\tan\theta = \underline{\hspace{2cm}}$

Holt Algebra 2

Reteach

LESSON 13-1

Right-Angle Trigonometry (continued)

The reciprocals of the sine, cosine, and tangent ratios are also trigonometric ratios.

The **cosecant** function (**csc** θ) is the reciprocal of the sine function.

$$\csc \theta = \frac{1}{\sin \theta} = \frac{\text{Hypotenuse}}{\text{Opposite}} = \frac{c}{a}$$

The **secant** function (**sec** θ) is the reciprocal of the cosine function.

$$\sec \theta = \frac{1}{\cos \theta} = \frac{\text{Hypotenuse}}{\text{Adjacent}} = \frac{c}{b}$$

The **cotangent** function (**cot** θ) is the reciprocal of the tangent function.

$$\cot \theta = \frac{1}{\tan \theta} = \frac{\text{Adjacent}}{\text{Opposite}} = \frac{b}{a}$$

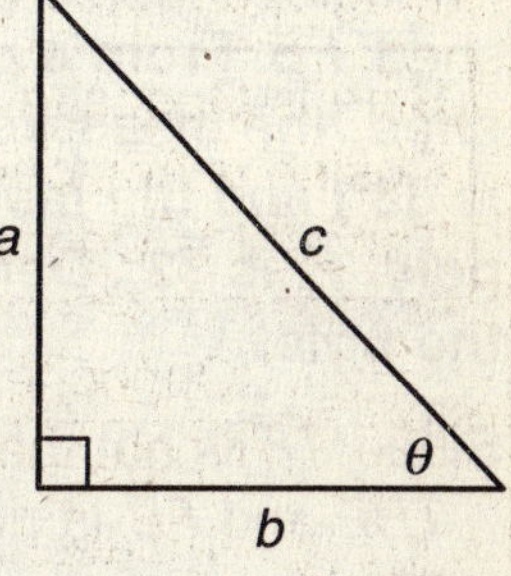

Use the reciprocal relationship of the ratios to find the values of the reciprocal trigonometric functions.

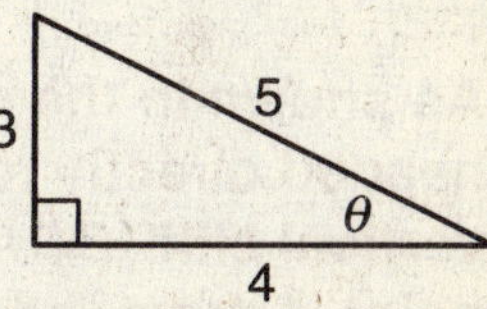

$$\sin \theta = \frac{3}{5} \qquad \csc \theta = \frac{1}{\sin \theta} = \frac{5}{3}$$

$$\cos \theta = \frac{4}{5} \qquad \sec \theta = \frac{1}{\cos \theta} = \frac{5}{4}$$

$$\tan \theta = \frac{3}{4} \qquad \cot \theta = \frac{1}{\tan \theta} = \frac{4}{3}$$

Find the values of the six trigonometric functions for θ.

4.

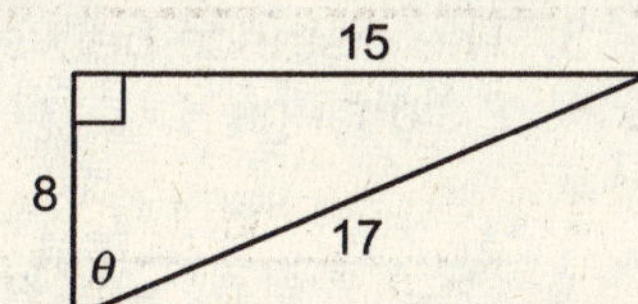

$\sin \theta = $ _____________

$\cos \theta = $ _____________

$\tan \theta = $ _____________

$\csc \theta = \dfrac{1}{\sin \theta} = $ _____________

$\sec \theta = \dfrac{1}{\cos \theta} = $ _____________

$\cot \theta = \dfrac{1}{\tan \theta} = $ _____________

5.

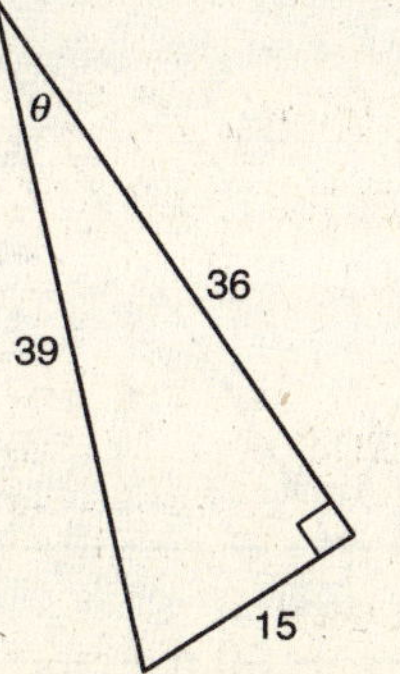

_____________   _____________

_____________   _____________

_____________   _____________

Holt Algebra 2

LESSON 13-1 **Challenge**
Twice Right

Trigonometry can be used to solve problems involving one or more right triangles. Different triangles can share common parts.

As shown in the diagram at right, a pole, $\overline{TF}$, is on the roof of a shed, $\overline{FB}$. From a point, P, on the ground 27 feet from the foot of the shed, the measure of the angle of elevation to the top of the pole, T, is 38°, and the measure of the angle of elevation to the foot of the pole, F, is 32°. Determine, to the nearest tenth of a foot, the height of the pole.

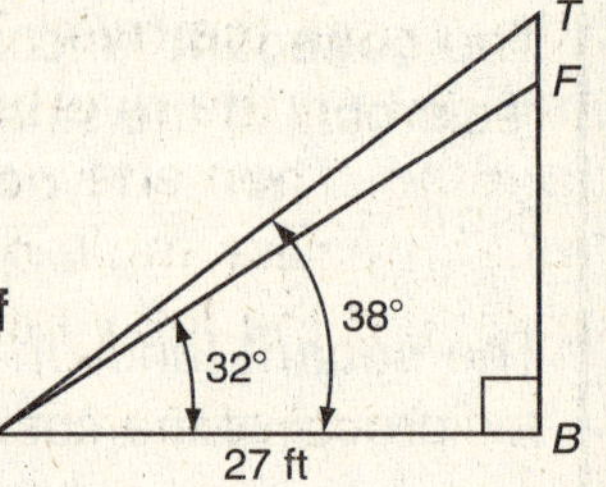

Since $\overline{TF}$ is not the side of a right triangle, first determine $\overline{TB}$ in right $\triangle TBP$ and $\overline{FB}$ in right $\triangle FBP$, and then subtract to determine $\overline{TF}$.

In right $\triangle TBP$:

$$\tan 38° = \frac{\text{opp.}}{\text{adj.}} = \frac{TB}{27}$$

$\overline{TB} = 27 \tan 38° \approx 21.09$

In right $\triangle FBP$:

$$\tan 32° = \frac{\text{opp.}}{\text{adj.}} = \frac{FB}{27}$$

$\overline{FB} = 27 \tan 32° \approx 16.87$

So, $\overline{TF} = \overline{TB} - \overline{FB} \approx 21.09 - 16.87 \approx 4.2$ feet.

As shown in the diagram at right, a ship is headed directly toward a coastline formed by a vertical cliff $(\overline{BC})$ that is 80 meters high. At point A, from the ship, the measure of the angle of elevation to the top of the cliff (B) is 10°. A few minutes later, at point D, the measure of the angle of elevation to the top of the cliff has increased to 20°. Determine the following to the nearest meter.

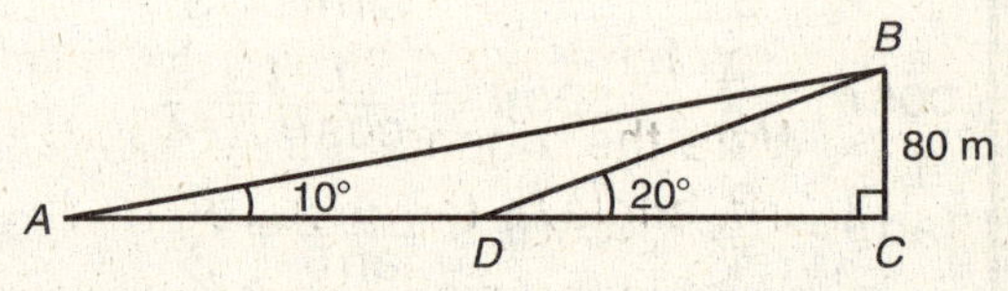

1. $\overline{DC}$

2. $\overline{AC}$

3. The distance between the 2 sightings

_______________   _______________   _______________

As shown in the diagram at right, the heights $(\overline{AB}$ and $\overline{CD})$ of buildings on opposite sides of a 90-foot-wide avenue are, respectively, 100 feet and 50 feet. From a point E on the avenue, the measure of the angle of elevation to B is 55°. Determine each of the following.

4. The distance from E to A to the nearest foot

____

5. The distance from E to D to the nearest foot

____

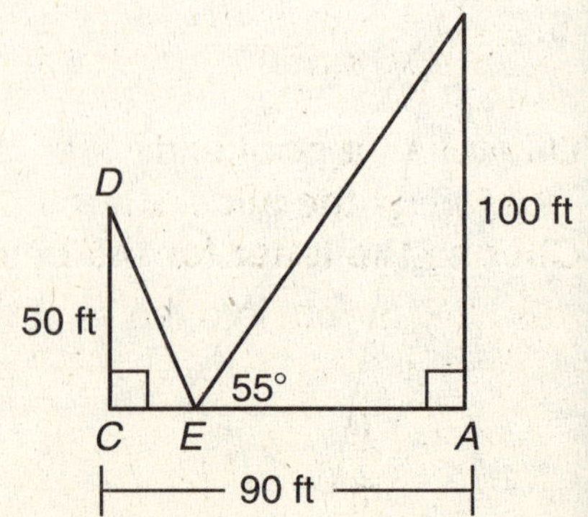

8

Holt Algebra 2

Problem Solving
Right-Angle Trigonometry

**Kayla is fishing near the ferry landing near her home.
She wonders how far the ferries actually travel when
they cross the river. To find out, she puts a fishing pole
upright on the riverbank directly across from the ferry
landing. Then she walks down the bank 180 yards and
measures an angle of 75°between the lines to her fishing
pole and the ferry landing on the opposite bank.**

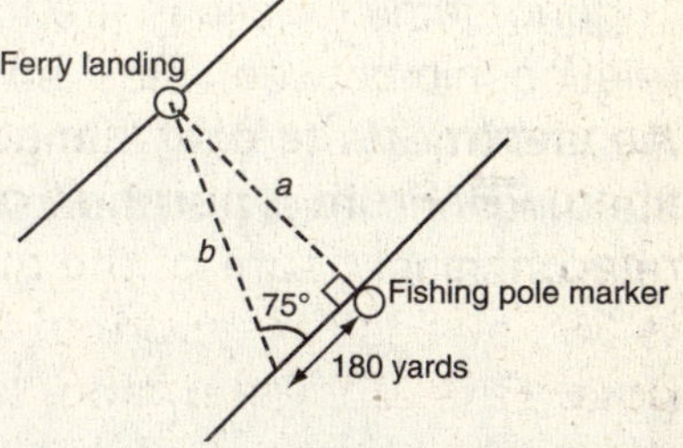

1. Find the distance directly across the river.

   a. On the diagram, name the side of the triangle
   that represents the distance Kayla wants to determine.
   ____

   b. Write a trigonometric function that relates the known
   distance and angle to the required distance.

   c. Find the distance to the nearest yard directly across
   the river.
   ____

2. Kayla walks another 200 yards down the riverbank. At this point the ferries seem to come
   straight at her at an angle of 60° to the line to her fishing pole. If the boats travel along this
   line to compensate for the current, how far do they travel?

   a. How far is she from her fishing pole now? _________________

   b. Write the trigonometric function that relates this distance
   and angle to the required distance, *c*.
   ____

   c. What is the distance that the ferries travel if they travel
   along this line?
   ____

**At a hot-air balloon festival, Luis watches a hot-air balloon rise from a
distance of 200 yards. Choose the letter for the best answer.**

3. From Luis's position, the balloon seems
   to hover at an angle of elevation of 50°.
   Which trigonometric function gives the
   height of the balloon, *h*?

   A $200\sin 50°$

   B $200\tan 50°$

   C $\dfrac{200}{\cos 50°}$

   D $\dfrac{200}{\cot 50°}$

4. After a short while, the balloon seems to
   hover at an angle of 75°. How high is it off
   the ground now?

   F 193 yd

   G 207 yd

   H 746 yd

   J 773 yd

**Olivia has a pool slide that makes an angle of 25° with the water.
The top of the slide stands 4.5 feet above the surface of the water.
Choose the letter for the best answer.**

5. How far out into the pool will the slide
   reach?

   A 2.1 ft

   B 5.0 ft

   C 7.6 ft

   D 9.7 ft

6. The slide makes a straight line into the
   water. How long is the slide?

   F 5.0 ft

   G 7.6 ft

   H 9.7 ft

   J 10.6 ft

Holt Algebra 2

Reading Strategy
13-1 Identify Relationships

Trigonometric functions are used to describe the relationship between side length and angle measurements of right triangles. Part of finding the functions involves understanding these relationships.

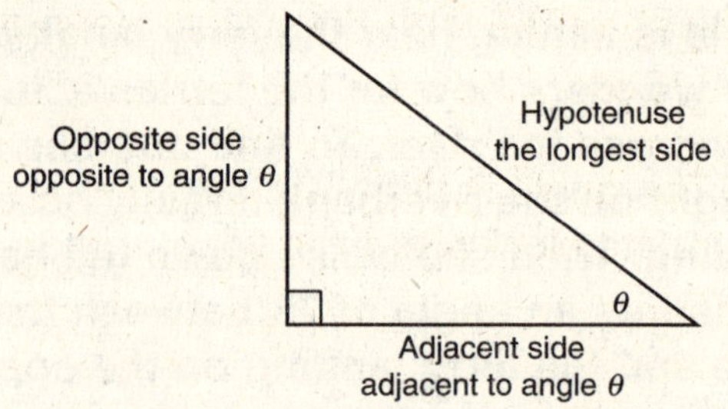

Triangle	sine	cosine	tangent
	$\sin\theta = \dfrac{\text{opposite}}{\text{hypotenuse}}$	$\cos\theta = \dfrac{\text{adjacent}}{\text{hypotenuse}}$	$\tan\theta = \dfrac{\text{opposite}}{\text{adjacent}}$
	$\sin\theta = \dfrac{\text{opp.}}{\text{hyp.}} = \dfrac{4}{5}$	$\cos\theta = \dfrac{\text{adj.}}{\text{hyp.}} = \dfrac{3}{5}$	$\tan\theta = \dfrac{\text{opp.}}{\text{adj.}} = \dfrac{4}{3}$

Answer each question.

1. For triangle *FGH*

   a. What is the length of the side opposite θ? _______________

   b. What is the length of the side adjacent to θ? _______________

   c. What is the length of the hypotenuse? _______________

   d. Write a fraction to represent cosine θ. _______________

   e. Write a fraction to represent tangent θ. _______________

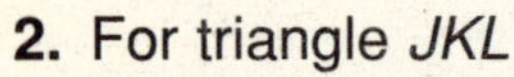

2. For triangle *JKL*

   a. What is the length of the side opposite θ? _______________

   b. What is the length of the side adjacent to θ? _______________

   c. What is the length of the hypotenuse? _______________

   d. Write a fraction to represent cosine θ. _______________

   e. Write a fraction to represent tangent θ. _______________

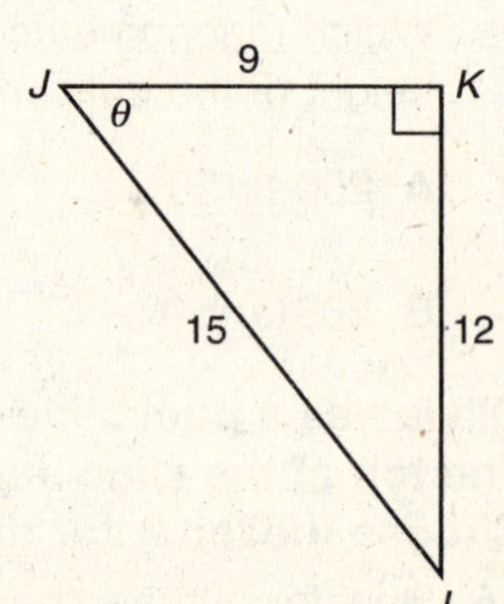

3. Compare the sine of the two acute angles in an isosceles right triangle. Explain.

   ____

   ____

Holt Algebra 2

LESSON 13-2 Practice A
Angles of Rotation

Draw an angle with the given measure in standard position.

1. $-390°$

The angle is negative, so rotate clockwise from $0°$.

2. $315°$

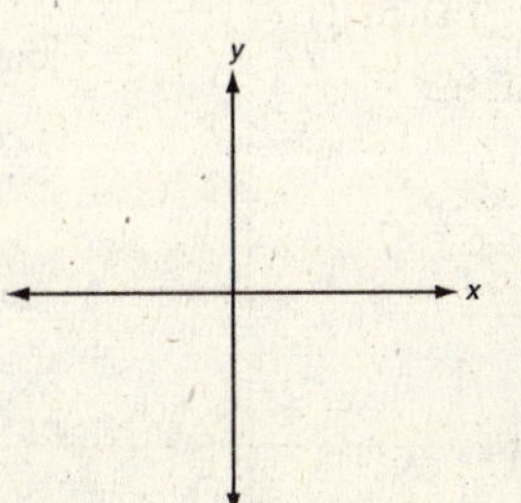

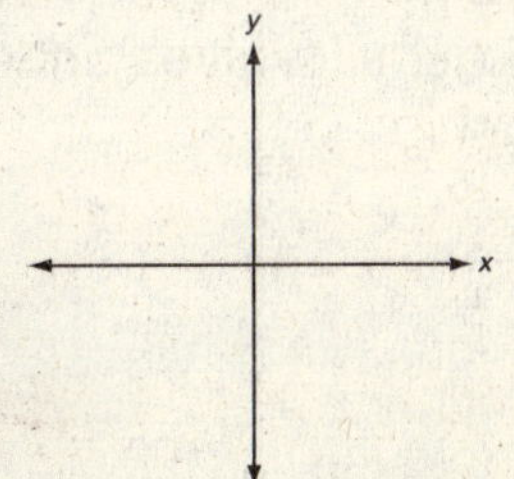

3. $-120°$

4. $240°$

5. $-585°$

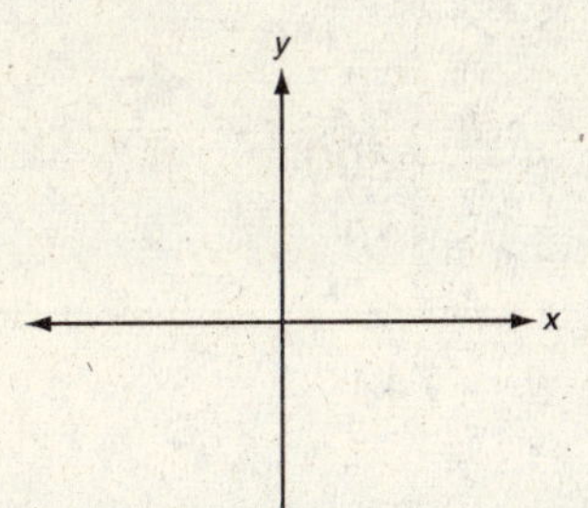

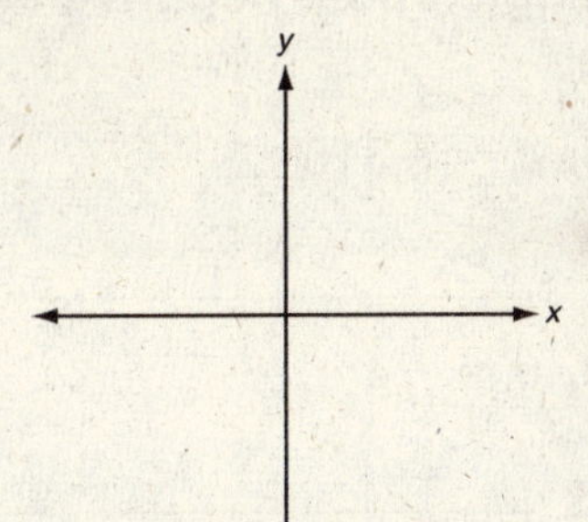

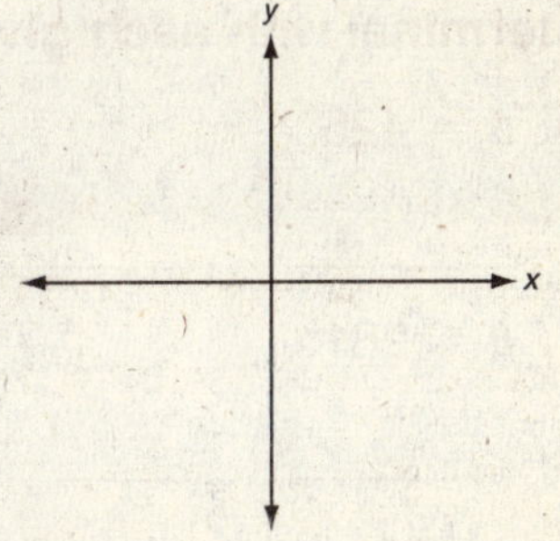

Find the measures of a positive angle and a negative angle that are coterminal with each given angle.

6. $\theta = 28°$

   a. Add $360°$. _______________

   b. Subtract $360°$. _______________

7. $\theta = 250°$

   a. Add $360°$. _______________

   b. Subtract $360°$. _______________

8. $\theta = 740°$

9. $\theta = -6°$

10. $\theta = -370°$

11. $\theta = 515°$

12. $\theta = -93°$

13. $\theta = 162°$

Find the measure of the reference angle for each given angle.

14. $\theta = 475°$

15. $\theta = 212°$

16. $\theta = -115°$

17. $\theta = 740°$

18. $\theta = -96°$

19. $\theta = -401°$

20. $\theta = 320°$

21. $\theta = -722°$

22. $\theta = 292°$

23. $\theta = -850°$

24. $\theta = 1000°$

25. $\theta = -1000°$

Holt Algebra 2

Practice B
Angles of Rotation

Draw an angle with the given measure in standard position.

1. $-420°$ 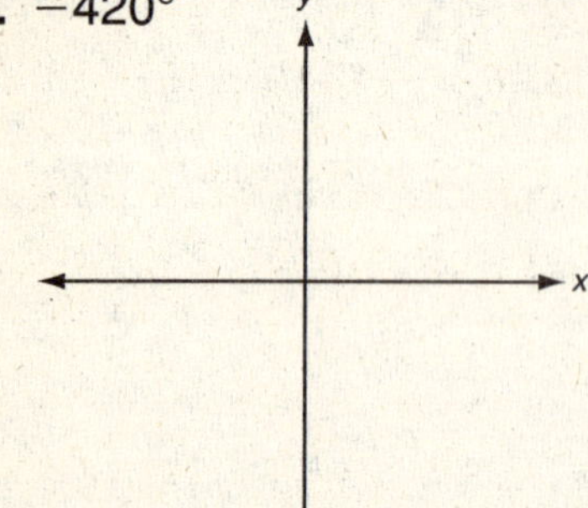

2. $405°$ 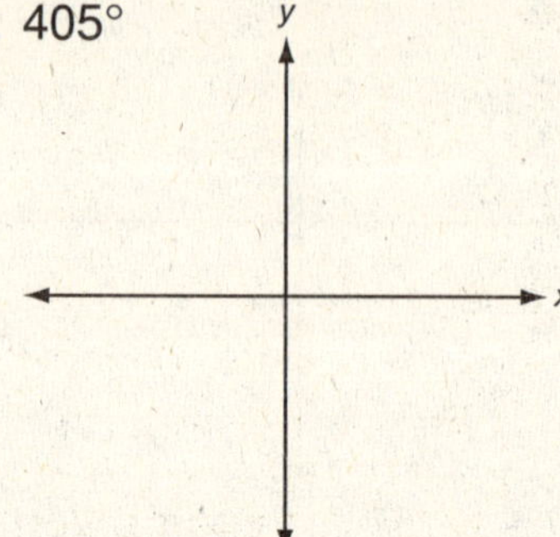

3. $-450°$ 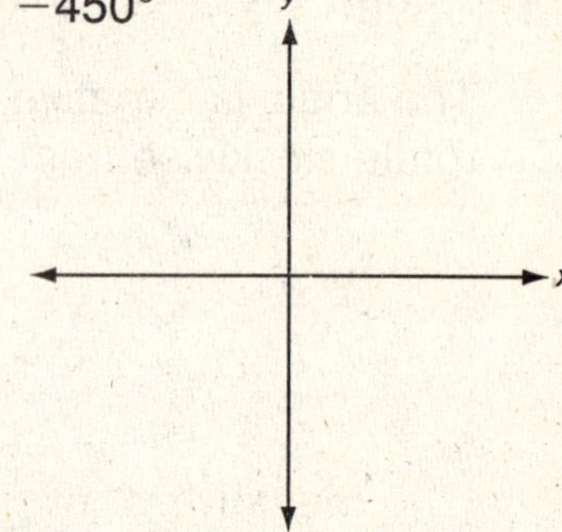

Find the measures of a positive angle and a negative angle that are coterminal with each given angle.

4. $\theta = 425°$

5. $\theta = -316°$

6. $\theta = -800°$

7. $\theta = 281°$

8. $\theta = -4°$

9. $\theta = 743°$

Find the measure of the reference angle for each given angle.

10. $\theta = 211°$

11. $\theta = -755°$

12. $\theta = -555°$

13. $\theta = 119°$

14. $\theta = -160°$

15. $\theta = 235°$

***P* is a point on the terminal side of θ in standard position. Find the exact value of the six trigonometric functions for θ.**

16. $P(-5, 5)$

17. $P(2, 9)$

18. $P(-7, -5)$

Solve.

19. A circus performer trots her pony into the ring. The pony circles the ring 22 times as the performer flips and turns on the pony's back. At the end of the act, the pony exits on the side of the ring opposite its point of entry. Through how many degrees does the pony trot during the entire act? __________

Holt Algebra 2

Practice C
LESSON 13-2 — *Angles of Rotation*

Find the measures of a positive angle and a negative angle that are coterminal with each given angle.

1. $\theta = 400°$

2. $\theta = -360°$

3. $\theta = -1010°$

4. $\theta = 567°$

5. $\theta = -164°$

6. $\theta = 358°$

Find the measure of the reference angle for each given angle.

7. $\theta = 504°$

8. $\theta = -388°$

9. $\theta = 991°$

10. $\theta = 486°$

11. $\theta = -920°$

12. $\theta = -1787°$

***P* is a point on the terminal side of θ in standard position. Find the exact value of the six trigonometric functions for θ.**

13. $P(-1, 14)$

14. $P(-8, -8)$

15. $P(9, -6)$

16. $P(10, 15)$

17. $P(-2, -1)$

18. $P(-12, 5)$

Solve.

19. A restaurant in the round rotates clockwise so diners can view the city. Fifty evenly-spaced window tables are numbered clockwise from 1 to 50. A waiter noted where Table 1 was at the beginning of his shift. At the end of his shift, the restaurant had made 4 complete rotations and Table 1 was then where Table 22 had been. Through how many degrees had the restaurant rotated during his shift? _________________

Holt Algebra 2

Reteach

Angles of Rotation

To draw an angle with a given measure, start with the initial side on the
x-axis and rotate the terminal side to show the angle measure.

- Rotate *counterclockwise* for a positive angle measure.

- Rotate *clockwise* for a negative angle measure.

> Label the angle measures on the axes to help locate the correct quadrant.

Draw a 210° angle.	Draw a −145° angle.
Start at the *x*-axis and rotate counterclockwise past 180° to 210°. Draw the terminal side. Label the angle.	Start at the *x*-axis and rotate clockwise past −90° to −145°. Draw the terminal side. Label the angle.

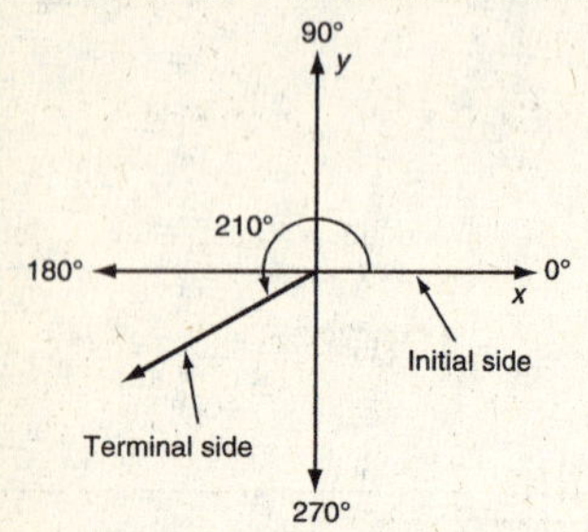

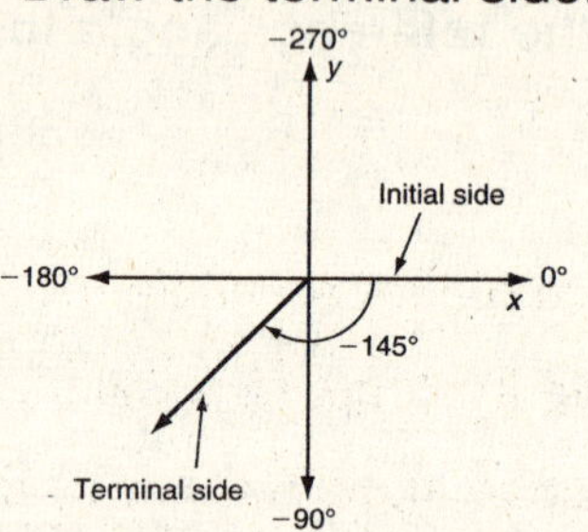

Coterminal angles are angles with the same initial and terminal sides.
You can find the measure of an angle that is coterminal with another angle
by adding 360° to or subtracting 360° from the angle.

> 360° is a complete rotation.

For $\theta = 65°$:

$$65° + 360° = 425°$$

$$65° - 360° = -295°$$

425° and −295° are coterminal with 65°.

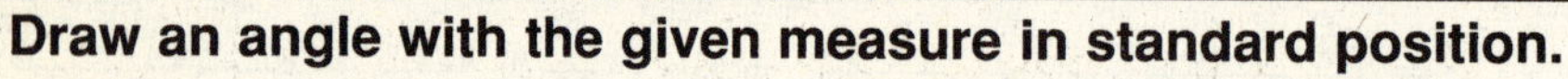

Draw an angle with the given measure in standard position.

1. 150°

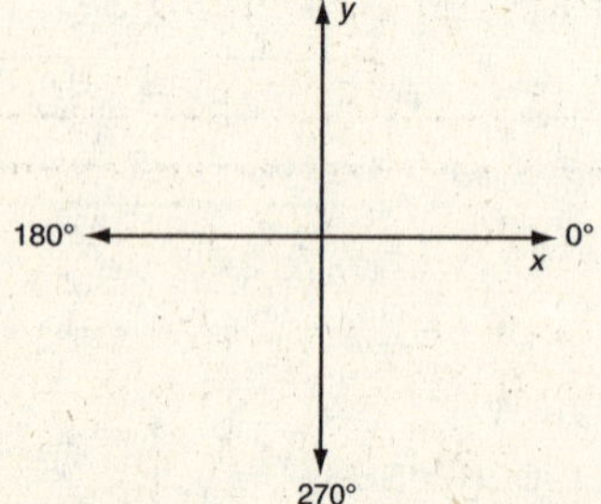

2. −330°

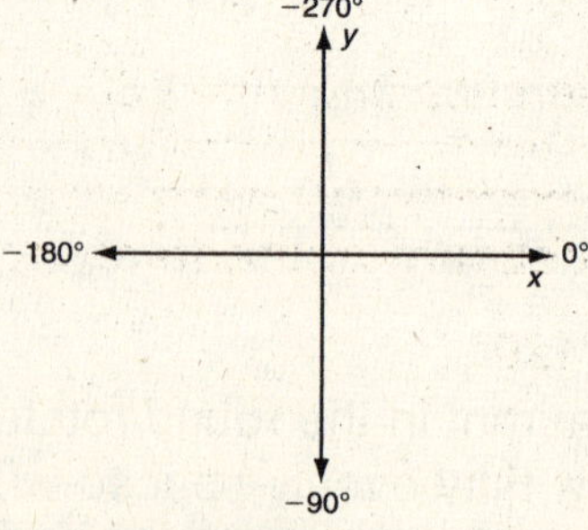

Find the measures of two angles that are coterminal with each angle.

3. 145°   **4.** −130°   **5.** 210°

$145° + 360° = $ __________   __________   __________

$145° - 360° = $ __________   __________   __________

Holt Algebra 2

LESSON 13-2 Reteach
Angles of Rotation (continued)

The **reference angle** for an angle θ in standard position is the positive acute angle formed by the terminal side of θ and the x-axis.

Denote the reference angle for θ as θ'. The diagram shows the location of the reference angle in each quadrant.

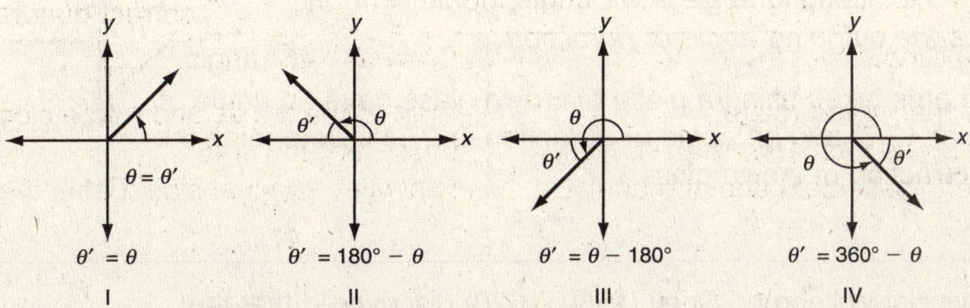

To find the reference angle for an angle θ:

- draw the angle in standard position,
- find the reference angle formed by the terminal side of θ and the x-axis,
- find the measure of θ'.

θ' is an acute angle so it is always less than 90°.

For $\theta = 130°$.

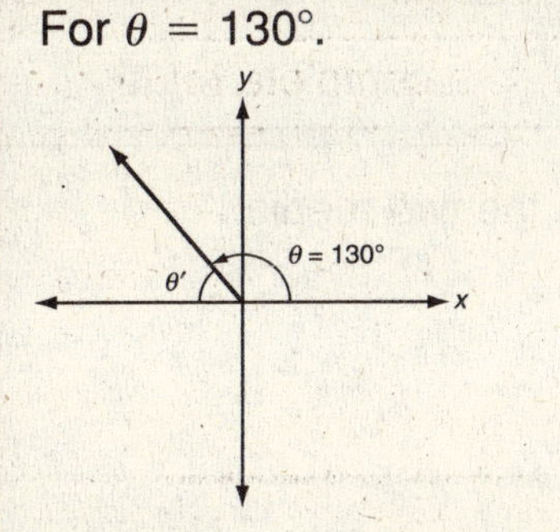

$\theta' = 180° - 130° = 50°$

The reference angle for 130° is 50°.

For $\theta = 225°$.

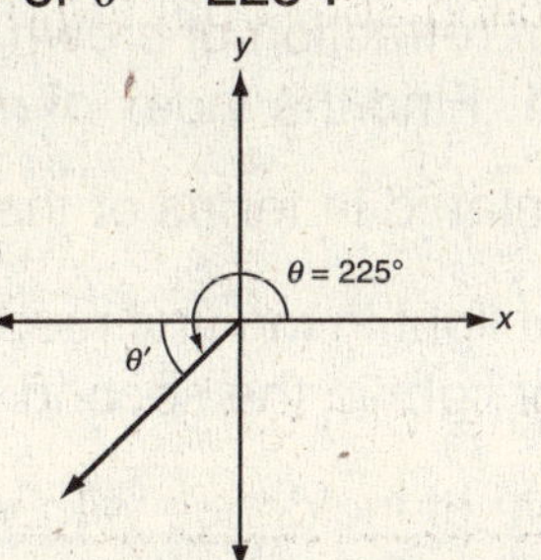

$\theta' = 225° - 180° = 45°$
The reference angle for 225° is 45°.

Draw θ. Then use the axis to find the measure of θ'.

Find the measure of the reference angle for each angle.

6. $\theta = -120°$

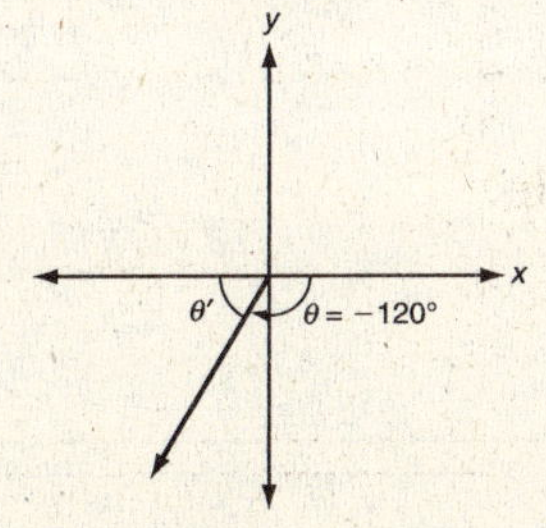

$\theta' = 180° - 120° =$ ______

7. $\theta = 315°$

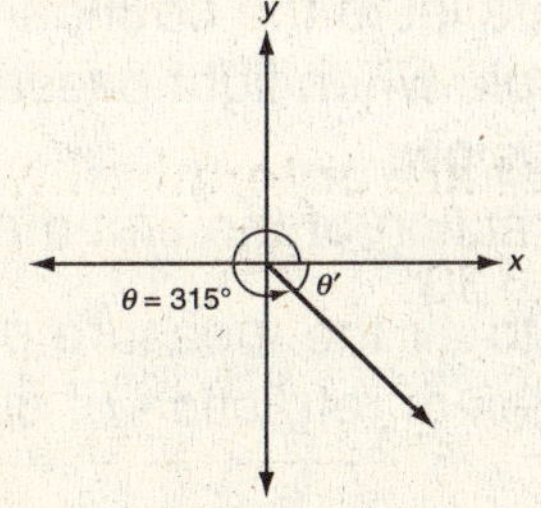

______

8. $\theta = 110°$

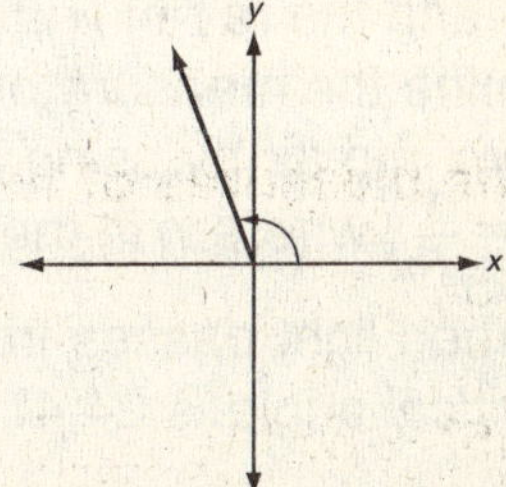

______

Holt Algebra 2

Challenge
Exploring the Behavior of Light

When a light ray passes from one medium, such as air, into another medium, such as glass, the light ray is bent, or *refracted*, at the boundary between the two media. The *index of refraction, n*, of a particular medium is given by the equation below.

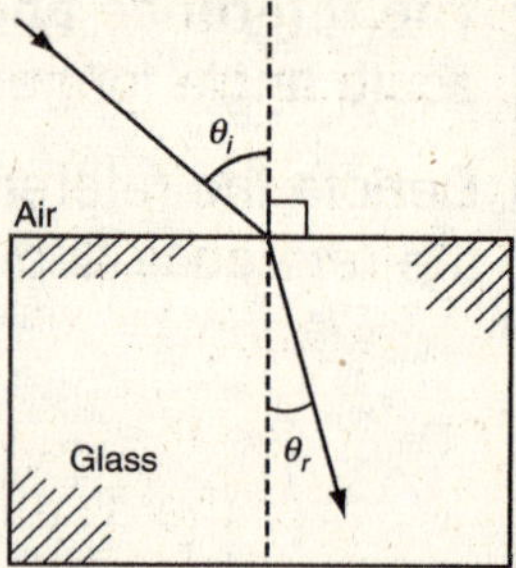

$$n = \frac{\sin\theta_i}{\sin\theta_r} \begin{cases} \theta_i \text{ is the incoming } \textit{angle of incidence} \text{ of the light ray} \\ \theta_r \textit{ is the} \text{ outgoing } \textit{angle of refraction} \end{cases}$$

1. Measurements taken using a piece of crown glass gave an angle of incidence of 43.7° and an angle of refraction of 27.1°. Determine the index of refraction of crown glass.

____

The Dutch physicist Willebrond Snell (1580–1626) discovered that the sine of the angle of incidence is proportional to the sine of the angle of refraction for any given interface between two media. As shown below, Snell's Law may be stated in terms of the indices of refraction of the media.

$$n_1 \sin\theta_i = n_2 \sin\theta_r \begin{cases} n_1 \text{ is the index of refraction of the first medium} \\ n_2 \text{ is the index of refraction of the second medium} \end{cases}$$

2. The measure of θ_i is 30° and the measure of θ_r for crown glass is 19.2°. Use the index of refraction of crown glass that you determined in Exercise 1. Find the index of refraction of air. __________

Snell's Law may be also be stated in terms of the velocity of light in the two media.

$$\frac{\sin\theta_i}{\sin\theta_r} = \frac{v_1}{v_2} \begin{cases} v_1 \text{ is the speed of light in the first medium} \\ v_2 \text{ is the speed of light in the second medium} \end{cases}$$

3. The measure of the angle of incidence with which a ray of light strikes crown glass is 30° and the measure of the angle of refraction is 19.2°. The speed of light in air is 3.0×10^8 meters per second. Find the speed of light in crown glass. __________

When light crosses a boundary and its speed increases in the second medium, the measure of the angle of refraction is greater than the measure of the angle of incidence. There is an angle of incidence, called the *critical angle* (θ_c), for which the corresponding angle of refraction is 90°, and the ray of light emerges parallel to the boundary. You can determine the measure of a critical angle when light passes from some medium into air by using the formula below.

$$\sin\theta_c = \frac{1}{n} \text{, where } n \text{ is the index of refraction of the first medium}$$

4. When light passes from diamond to air the measure of the critical angle is 24.4°. Find the index of refraction of diamond. __________

Holt Algebra 2

LESSON 13-2 — **Problem Solving**
Angles of Rotation

Isabelle and Karl agreed to meet at the rotating restaurant at the top of a tower in the town center. The restaurant makes one full rotation each hour.

1. Isabelle waits for Karl, who arrives 30 minutes later. Through how many degrees does the restaurant rotate between the time that Isabelle arrives and the time that Karl arrives? _______________

2. Since the restaurant is busy, they don't get menus for another ten minutes. How much farther has the restaurant rotated? _______________

3. By the time they are served dinner, the restaurant has rotated to an angle 30° short of its orientation when Isabelle arrived.

   a. Write an expression for the length of time that Isabelle has been there. _______________

   b. How long has she been there? _______________

   c. How long has Karl been there? _______________

   d. How far has the restaurant rotated since Karl arrived? _______________

4. When their bill comes, it includes a note that the restaurant has rotated 840° since Isabelle arrived.

   a. How long has it been since they were served dinner? _______________

   b. How many rotations has the restaurant made since Isabelle arrived? _______________

   c. How far is the restaurant from its orientation when Karl arrived? _______________

5. On their way out, they stop and look at a map. A museum is located at a point northeast of the tower with coordinates (3, 2).

   a. Write the trigonometric function for the angle that the line from the tower to the museum makes with a line due north from the tower. _______________

   b. Isabelle says that the museum is exactly 4 kilometers from the tower. How much farther north is the museum than the tower? _______________

An advertisement on a kiosk near the bus stop rotates through 765° while Aaron waits for his bus. Choose the letter for the best answer.

6. What is the difference in the orientation of the advertisement between when Aaron arrived at the bus stop and when the bus came?

   A 45°

   B 90°

   C 315°

   D 405°

7. If the advertisement rotates at a rate of one rotation every 10 minutes, how long does Aaron wait for his bus?

   F Less than 2 min

   G Between 2 min and 10 min

   H Between 10 min and 20 min

   J More than 20 min

Holt Algebra 2

LESSON 13-2 Reading Strategy
Understand Vocabulary

There are different terms that are used to describe an angle of rotation on the coordinate plane.

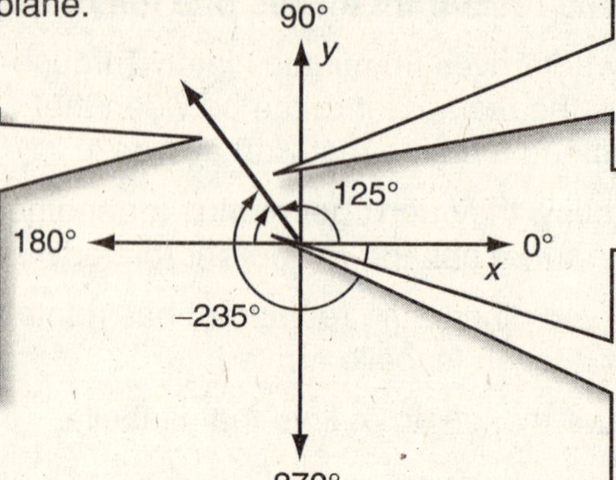

This angle is in **standard position**: Its vertex is at the origin and one ray is on the positive *x*-axis. This is its **terminal side**. It measures 125°.

Coterminal angles are angles in standard position with the same terminal side. The angle measuring −235° is coterminal with the angle measuring 125°.

The **reference angle** is the positive acute angle formed by the terminal side of an angle and the *x*-axis. This reference angle is 55°.

Answer each question.

1. An angle has a counterclockwise angle of rotation and the terminal side lies in the third quadrant. What do you know about this angle?

____

2. An angle has a clockwise angle of rotation and the terminal side lies in the first quadrant. What do you know about this angle?

____

3. **a.** Draw a 245° angle in standard position.

   b. What is the measure of a coterminal angle?

   ____

   c. What is the measure of the reference angle?

   ____

4. **a.** Draw a −200° angle in standard position.

   b. What is the measure of a coterminal angle?

   ____

   c. What is the measure of the reference angle?

   ____

5. Do all angles have both positive and negative coterminal angles? Explain.

____

____

6. Are reference angles always positive? Explain.

____

____

Holt Algebra 2

LESSON 13-3

Practice A
The Unit Circle

Convert each measure from degrees to radians or from radians to degrees.

1. $60°$

$60° \left(\dfrac{\pi \text{ radians}}{180°} \right) =$ _____________

2. $-\dfrac{2\pi}{5}$

$\left(-\dfrac{2\pi}{5} \right) \left(\dfrac{180°}{\pi \text{ radians}} \right) =$ _____________

3. $\dfrac{5\pi}{6}$

4. $315°$

5. $-\dfrac{3\pi}{4}$

6. $-105°$

7. $\dfrac{4\pi}{3}$

8. $-\dfrac{\pi}{6}$

9. $300°$

10. $-10°$

11. $\dfrac{16\pi}{9}$

Find the exact value of each trigonometric function. Use the unit circle.

12. $\sin 60°$

   a. At what point on the unit circle does the angle terminate? _____________

   b. Use $\sin \theta = y$. _____________

13. $\cos \dfrac{5\pi}{3}$

14. $\tan 225°$

15. $\tan \pi$

16. $\sin 330°$

17. $\cos 150°$

18. $\tan 240°$

Solve.

19. John is adding a curved edge to the landscaping in front of the high school. The curve is an arc of a circle with a radius of 1600 feet. The central angle that intercepts the curve measures $\dfrac{\pi}{8}$ radians. Find the length of the curve to the nearest foot.

____

Holt Algebra 2

<table><tr><td>LESSON
13-3</td><td>

Practice B
The Unit Circle
</td></tr></table>

Convert each measure from degrees to radians or from radians to degrees.

1. $\dfrac{5\pi}{12}$

2. $215°$

3. $-\dfrac{29\pi}{18}$

4. $-180°$

5. $\dfrac{5\pi}{3}$

6. $-\dfrac{7\pi}{6}$

7. $400°$

8. $\dfrac{3\pi}{10}$

9. $35°$

Use the unit circle to find the exact value of each trigonometric function.

10. $\cos\dfrac{2\pi}{3}$

11. $\tan\dfrac{5\pi}{4}$

12. $\tan\dfrac{5\pi}{6}$

13. $\sin 315°$

14. $\cos 225°$

15. $\tan 60°$

Use a reference angle to find the exact value of the sine, cosine, and tangent of each angle.

16. $150°$

17. $-225°$

18. $-300°$

19. $\dfrac{11\pi}{6}$

20. $-\dfrac{2\pi}{3}$

21. $\dfrac{5\pi}{4}$

Solve.

22. San Antonio, Texas, is located about $30°$ north of the equator. If Earth's radius is about 3959 miles, approximately how many miles is San Antonio from the equator?

Holt Algebra 2

Name ____ Date ___________ Class _____________

Practice C
The Unit Circle

Convert each measure from degrees to radians or from radians to degrees.

1. $-\dfrac{3\pi}{2}$

____

2. $450°$

____

3. $\dfrac{5\pi}{18}$

____

4. $-200°$

____

5. $\dfrac{7\pi}{4}$

____

6. $-\dfrac{11\pi}{6}$

____

7. $350°$

____

8. $\dfrac{7\pi}{20}$

____

9. $12°$

____

10. $\dfrac{13\pi}{10}$

____

11. $222°$

____

12. $-105°$

____

Find the exact value of the sine, cosine, and tangent of each angle.

13. $330°$

____

14. $\dfrac{7\pi}{4}$

____

15. $240°$

____

16. $\dfrac{5\pi}{6}$

____

17. $225°$

____

18. $120°$

____

19. $45°$

____

20. $-\pi$

____

21. $-\dfrac{5\pi}{6}$

____

22. $-\dfrac{\pi}{4}$

____

23. $-\dfrac{\pi}{3}$

____

24. $135°$

____

Solve.

25. A pendulum is 18 feet long. Its central angle is 44°. The pendulum makes one back and forth swing every 12 seconds. To the nearest foot, how far does the pendulum swing each minute?

____

21

Holt Algebra 2

LESSON 13-3 Reteach
The Unit Circle

Radians are a real number measure of rotation.
To convert between radians and degrees, use the following identity.

$$\pi \text{ radians} = 180°$$

To convert from degrees to radians, solve the identity for 1 degree.	To convert from radians to degrees, solve the identity for 1 radian.
$1 \text{ degree} = \dfrac{\pi \text{ radians}}{180°}$	$1 \text{ radian} = \dfrac{180°}{\pi \text{ radians}}$

Convert 60° to radians.

$$60° = 60°\left(\frac{\pi \text{ radians}}{180°^{\,3}}\right) = \frac{\pi}{3} \text{ radians}$$

> Use dimensional analysis to help. Notice that the degrees cancel so the remaining unit is radians.

Convert $\dfrac{5\pi}{4}$ radians to degrees.

$$\frac{5\pi}{4} \text{ radians} = \left(\frac{5\pi}{4} \text{ radians}\right)\left(\frac{^{45}180°}{\pi \text{ radians}}\right) = 225°$$

> The radians cancel so the remaining unit is degrees.

Convert each measure from degrees to radians.

1. −45°

$$-45° = -45°\left(\frac{\pi \text{ radians}}{180°}\right)$$

2. 150°

$$150° = 150°\left(\frac{\pi \text{ radians}}{180°}\right)$$

3. 210°

4. −120°

Convert each measure from radians to degrees.

5. $\dfrac{4\pi}{3}$ radians

$$\frac{4\pi}{3} \text{ radians} = \left(\frac{4\pi}{3} \text{ radians}\right)\left(\frac{180°}{\pi \text{ radians}}\right)$$

6. $-\dfrac{3\pi}{2}$ radians

$$-\frac{3\pi}{2} \text{ radians} = \left(-\frac{3\pi}{2} \text{ radians}\right)\left(\frac{180°}{\pi \text{ radians}}\right)$$

7. $\dfrac{\pi}{6}$ radians

8. $\dfrac{5\pi}{3}$ radians

Holt Algebra 2

Reteach
LESSON 13-3 *The Unit Circle (continued)*

Use a reference angle to find the exact value of the sine, cosine, and tangent of an angle in any quadrant.

Find the value of the trigonometric functions of 150°.

Step 1 Sketch the angle.
Find the measure of the reference angle.
$180° - 150° = 30°$

Step 2 Draw the triangle with the reference angle.
Label the sides with their lengths.

Step 3 Find the sine, cosine, and tangent of 30°.
$$\sin 30° = \frac{1}{2}$$
$$\cos 30° = \frac{\sqrt{3}}{2}$$
$$\tan 30° = \frac{1}{\sqrt{3}}$$

Step 4 Adjust the signs for 150°.
$$\sin 150° = \frac{1}{2}$$
$$\cos 150° = -\frac{\sqrt{3}}{2}$$
$$\tan 150° = -\frac{1}{\sqrt{3}}$$

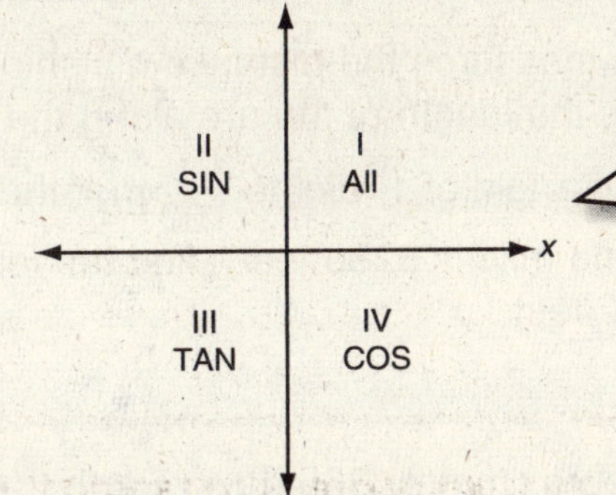

The diagram shows the quadrants in which each trigonometric function is positive.

Complete to find the exact value of the sine, cosine, and tangent of 315°.

9. Find the measure of the reference angle. _______________________

10. Find the sine, cosine, and tangent of the reference angle.

____

____

____

11. Adjust the signs to find the sine, cosine, and tangent of 315°.

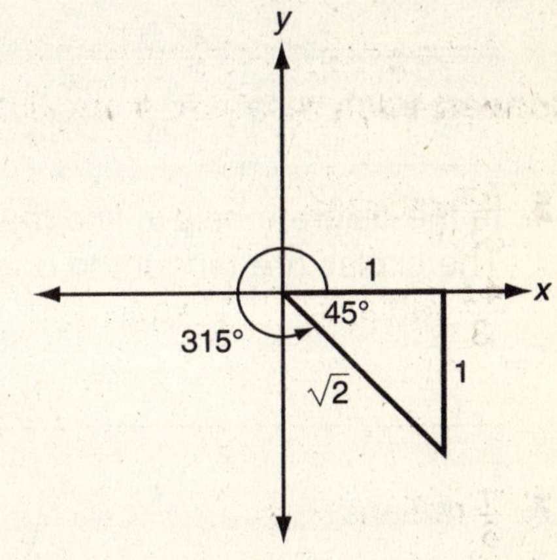

____

____

Holt Algebra 2

Challenge
LESSON 13-3 *A Radian as a Measure of Length*

Radians can be useful in application problems because of the close connection between radian measure for angles and arc length in a circle. Consider the formula for finding the length of an arc: $s = r\theta$. The variables s and r are both lengths, s is the length of the arc and r is the radius of the circle, and θ is an angle measured in radians. A radian is sometimes considered a unitless measure since it measures the distance around a circle using the radius as the measure of length. For instance, an angle of 2 radians is related to an arc of 2 radii around the circle. It makes no difference the size of the circle; an arc of 2 radii around the circle still produces the same angle.

Solve the following application problems by using the formula $s = r\theta$.

1. A captain at sea measures the distance his ship travels in nautical miles. A nautical mile is the length of the arc along the surface of Earth that is intercepted by an angle of 1 minute $\left(\frac{1}{60}\right)^{\circ}$. If the radius of Earth is 3960 miles (1 land mile is 5280 feet), find the length of a nautical mile to the nearest 10 feet.

   ____

2. One year has approximately 365 days. Earth travels around the sun in an approximately circular orbit with a radius of 93 million miles. Find the distance Earth travels in its orbit each day. Calculate the speed of Earth in its orbit in miles per hour. Round answers to the nearest whole number.

   ____

3. Use the formula for the area of a circle, $A = \pi r^2$, and derive the formula for the area of a sector using only the variables r and θ. Assume that θ is in radians.

   ____

   ____

   ____

4. In the diagram below, find the area of the shaded sector. The circles are tangent to one another as shown in the diagram. ________________

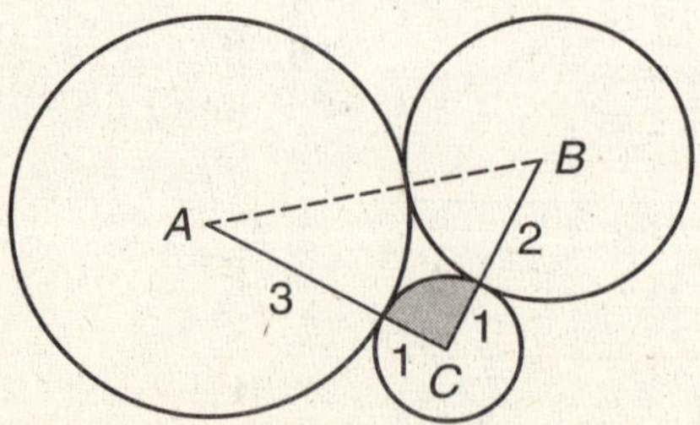

Holt Algebra 2

Problem Solving
The Unit Circle

**Gabe is spending two weeks on an archaeological dig.
He finds a fragment of a circular plate that his leader
thinks may be valuable. The arc length of the fragment
is about $\frac{1}{6}$ the circumference of the original complete
plate and measures 1.65 inches.**

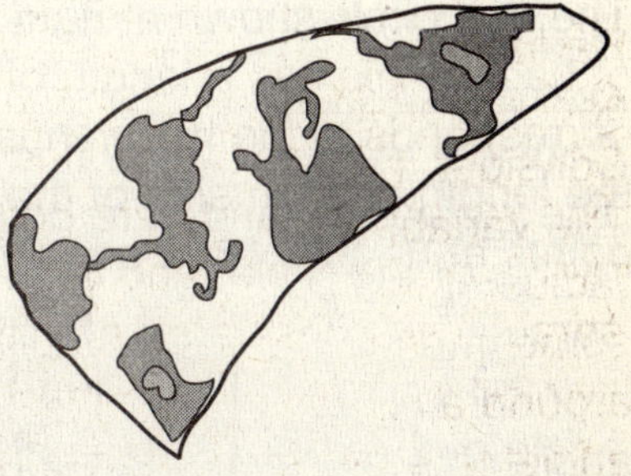

1. A similar plate found earlier has a diameter of 3.14 inches.
   Could Gabe's fragment match this plate?

   a. Write an expression for the radius, *r*, of the earlier plate.   _______________

   b. What is the measure, in radians, of a central angle, θ,
      that intercepts an arc that is $\frac{1}{6}$ the length of the
      circumference of a circle?   _______________

   c. Write an expression for the arc length, *S*, intercepted
      by this central angle.   _______________

   d. How long would the arc length of a fragment be if it were $\frac{1}{6}$
      the circumference of the plate?   _______________

   e. Could Gabe's plate be a matching plate? Explain.

   ____

   ____

2. Toby finds another fragment of arc length 2.48 inches. What
   fraction of the outer edge of Gabe's plate would it be if this
   fragment were part of Gabe's plate?   _______________

**The diameter of a merry-go-round at the playground is 12 feet. Elijah
stands on the edge and his sister pushes him around. Choose the
letter for the best answer**

3. How far does Elijah travel if he moves
   through an angle of $\frac{5\pi}{4}$ radians?

   A 12.0 ft   C 23.6 ft

   B 15.1 ft   D 47.1 ft

4. Through what angle does Elijah move if
   he travels a distance of 80 feet around
   the circumference?

   F $\frac{40}{3}\pi$ radians   H $\frac{40}{3}$ radians

   G $\frac{80}{3}$ radians   J $\frac{20}{3}$ radians

**Virgil sets his boat on a 1000-yard course keeping a constant
distance from a rocky outcrop. Choose the letter for the best answer.**

5. If Virgil keeps a distance of 200 yards,
   through what angle does he travel?

   A 5π radians   C 10 radians

   B 5 radians   D 10π radians

6. If Virgil keeps a distance of 500 yards,
   what fraction of the circumference of a
   circle does he cover?

   F $\frac{1}{\pi}$ H $\frac{3}{4\pi}$

   G $\frac{\pi}{3}$ J $\frac{3\pi}{4}$

Holt Algebra 2

LESSON 13-3 Reading Strategy
Use a Visual Map

The unit circle shown at right has a radius of 1 unit. So the terminal side of each angle has a length of 1 unit. As shown in the diagram below, that corresponds to the hypotenuse of a right triangle. The ordered pair shows the length of each side of the triangle. Notice the ordered pairs for the points at 0°, 90°, 180°, and 270°.

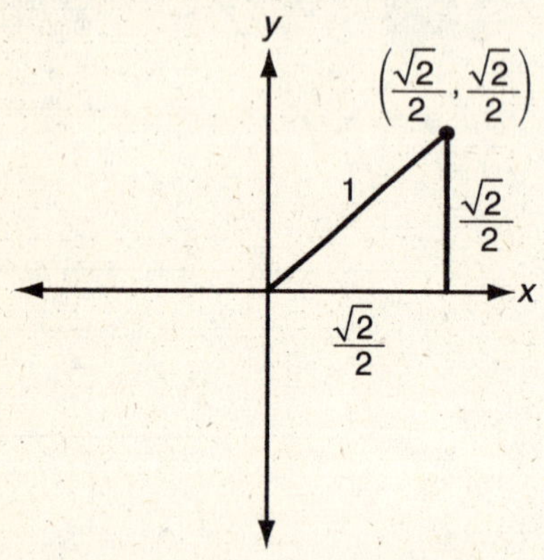

The Unit Circle

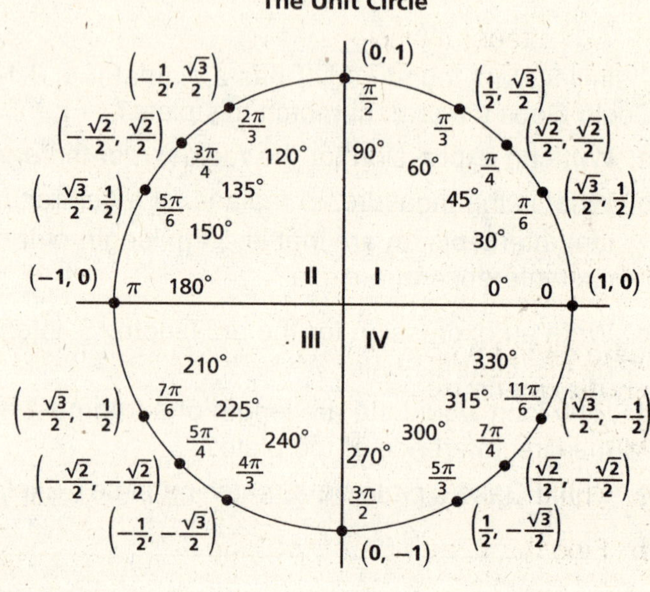

On the circle, for every value of θ,

$$\sin\theta = \frac{y}{r} = \frac{y}{1} = y$$

$$\cos\theta = \frac{x}{r} = \frac{x}{1} = x$$

$$\tan\theta = \frac{y}{x}$$

Use the unit circle to answer each question.

1. Express 360° in radians.

2. What is sin360°?

3. What is cos360°?

_________________________   _________________________   _________________________

4. a. In which quadrant does the terminal side of a 150° angle lie?

b. Express 150° in radians.

c. What point does the terminal side of an angle of 150° pass through on the unit circle?

d. Which coordinate of the ordered pair represents sin150°?

e. Which coordinate of the ordered pair represents cos150°?

f. Write an expression for tan150° and simplify.

Holt Algebra 2

LESSON 13-4 Practice A
Inverses of Trigonometric Functions

Find all possible values of each expression.

1. $\cos^{-1}\left(-\dfrac{\sqrt{3}}{2}\right)$

    a. Since $\cos\theta = x$, look at the x-coordinates on the unit circle.
    Find the points where the x-coordinate is $-\dfrac{\sqrt{3}}{2}$. ________________

    b. Add integer multiples of $2\pi n$. ________________

2. $\sin^{-1}\left(\dfrac{1}{2}\right)$ 3. $\cos^{-1}\left(\dfrac{1}{2}\right)$ 4. $\tan^{-1}1$

5. $\sin^{-1}\left(\dfrac{\sqrt{3}}{2}\right)$ ________________   6. $\cos^{-1}\left(-\dfrac{1}{2}\right)$ ________________

Evaluate each inverse trigonometric function. Give your answer in both radians and degrees.

7. $\text{Sin}^{-1}0$

    a. The inverse sine function is restricted to which quadrants? ________________

    b. Find the value of θ whose sine is 0. ________________

8. $\text{Sin}^{-1}\left(-\dfrac{\sqrt{2}}{2}\right)$ 9. $\text{Tan}^{-1}(\sqrt{3})$ 10. $\text{Cos}^{-1}\left(\dfrac{\sqrt{3}}{2}\right)$

11. $\text{Tan}^{-1}(-1)$ 12. $\text{Cos}^{-1}(-1)$ 13. $\text{Sin}^{-1}(-1)$

Solve each equation to the nearest tenth. Use the given restrictions.

14. $\sin\theta = 0.95$, for $-90° \le \theta \le 90°$ 15. $\cos\theta = 0.2$, for $270° \le \theta \le 360°$

Solve.

16. The pilot of a plane wants to fly 800 miles east and 155 miles north of the airport. To the nearest degree, in what direction should he head?

Holt Algebra 2

LESSON 13-4 Practice B
Inverses of Trigonometric Functions

Find all possible values of each expression.

1. $\sin^{-1}\left(-\dfrac{\sqrt{3}}{2}\right)$

2. $\cos^{-1}\left(-\dfrac{1}{2}\right)$

3. $\tan^{-1}0$

____

4. $\sin^{-1}\left(-\dfrac{\sqrt{2}}{2}\right)$

5. $\cos^{-1}\left(-\dfrac{\sqrt{2}}{2}\right)$

6. $\tan^{-1}\left(\dfrac{\sqrt{3}}{3}\right)$

____

Evaluate each inverse trigonometric function. Give your answer in both radians and degrees.

7. $\mathrm{Sin}^{-1}(-1)$

8. $\mathrm{Tan}^{-1}(-\sqrt{3})$

9. $\mathrm{Cos}^{-1}1$

____

10. $\mathrm{Sin}^{-1}\left(\dfrac{\sqrt{3}}{2}\right)$

11. $\mathrm{Tan}^{-1}\left(-\dfrac{\sqrt{3}}{3}\right)$

12. $\mathrm{Cos}^{-1}\left(\dfrac{\sqrt{2}}{2}\right)$

____

Solve each equation to the nearest tenth. Use the given restrictions.

13. $\sin\theta = 0.45$, for $0° < \theta < 90°$

14. $\sin\theta = 0.801$, for $90° < \theta < 270°$

____

15. $\tan\theta = 2.42$, for $180° < \theta < 360°$

16. $\cos\theta = -0.334$, for $0° < \theta < 180°$

____

17. $\cos\theta = -0.181$, for $180° < \theta < 360°$

18. $\tan\theta = -10$, for $90° < \theta < 270°$

____

Solve.

19. A 21-foot ladder is leaning against a building. The base of the ladder is 7 feet from the base of a building. To the nearest degree, what is the measure of the angle that the ladder makes with the ground?

____

Holt Algebra 2

Practice C
Inverses of Trigonometric Functions

Find all possible values of each expression.

1. $\sin^{-1}\left(\dfrac{\sqrt{3}}{2}\right)$

2. $\cos^{-1}\left(\dfrac{\sqrt{2}}{2}\right)$

3. $\tan^{-1}(-\sqrt{3})$

4. $\cos^{-1}\left(-\dfrac{\sqrt{2}}{2}\right)$

5. $\sin^{-1}\left(\dfrac{1}{2}\right)$

6. $\tan^{-1}(-1)$

Evaluate each inverse trigonometric function. Give your answer in both radians and degrees.

7. $\text{Sin}^{-1}\left(-\dfrac{\sqrt{3}}{2}\right)$

8. $\text{Cos}^{-1}\left(-\dfrac{1}{2}\right)$

9. $\text{Tan}^{-1}\left(\dfrac{\sqrt{3}}{3}\right)$

10. $\text{Sin}^{-1}(1)$

11. $\text{Tan}^{-1}(\sqrt{3})$

12. $\text{Cos}^{-1}0$

Solve each equation to the nearest tenth. Use the given restrictions.

13. $\sin\theta = -0.204$, for $90° < \theta < 270°$

14. $\tan\theta = 8.055$, for $90° < \theta < 270°$

15. $\cos\theta = 0.778$, for $180° < \theta < 360°$

16. $\tan\theta = -4$, for $0° < \theta < 180°$

17. $\sin\theta = 0.094$, for $90° < \theta < 270°$

18. $\cos\theta = -0.555$, for $90° < \theta < 270°$

Solve.

19. A gutter at the edge of a roof drops 2 inches for every
30 feet of length. To the nearest tenth of a degree, what
is the measure of the angle that the gutter makes with
the roof?

Holt Algebra 2

LESSON 13-4 Reteach
Inverses of Trigonometric Functions

The trigonometric functions have inverse relations.

Trigonometric Function	Inverse Relation
$\sin\theta = a$	$\sin^{-1}a = \theta$
$\cos\theta = a$	$\cos^{-1}a = \theta$
$\tan\theta = a$	$\tan^{-1}a = \theta$

Read "the inverse sine of a."

Read "the inverse cosine of a."

Read "the inverse tangent of a."

Evaluate the inverse of a trigonometric function.

Step 1 Each inverse trigonometric relation has multiple values.

Think, "What reference angle has the given trigonometric value?"

Find the two angles between 0 and 2π radians that have the given value for the trigonometric function.

Step 2 Add $(2\pi)n$ to each angle to represent all the coterminal angles, where n is an integer.

Find all possible values of $\cos^{-1}\dfrac{\sqrt{3}}{2}$.

Think: What reference angle has a cosine of $\dfrac{\sqrt{3}}{2}$?

Step 1 The value of the cosine is positive.

Cosine is positive in Quadrants I and IV.

In Quadrant I, the reference angle is the angle.

$\cos\dfrac{\pi}{6} = \dfrac{\sqrt{3}}{2}$ so $\cos^{-1}\dfrac{\sqrt{3}}{2} = \dfrac{\pi}{6}$

To find the angle in Quadrant IV, subtract the reference angle from 2π.

$2\pi - \dfrac{\pi}{6} = \dfrac{11\pi}{6}$

Step 2 Add $(2\pi)n$ to each angle.

$\dfrac{\pi}{6} + 2\pi n$ or $\dfrac{11\pi}{6} + 2\pi n$

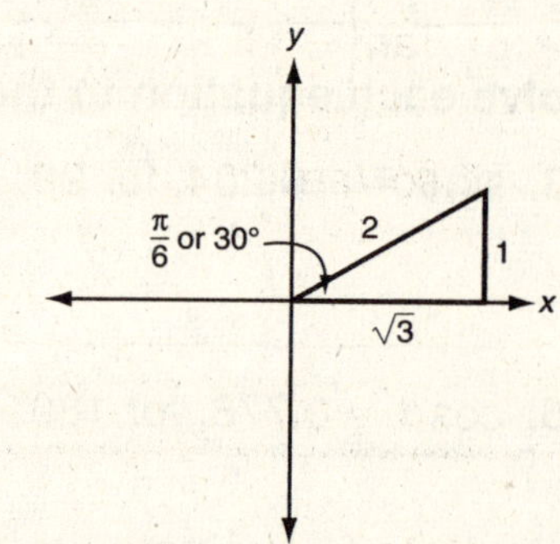

Complete to find all possible values of $\tan^{-1}\sqrt{3}$.

1. In which quadrants is tangent positive?

____

2. Name one angle that has tangent of $\sqrt{3}$.

____

3. Name another angle that has tangent of $\sqrt{3}$.

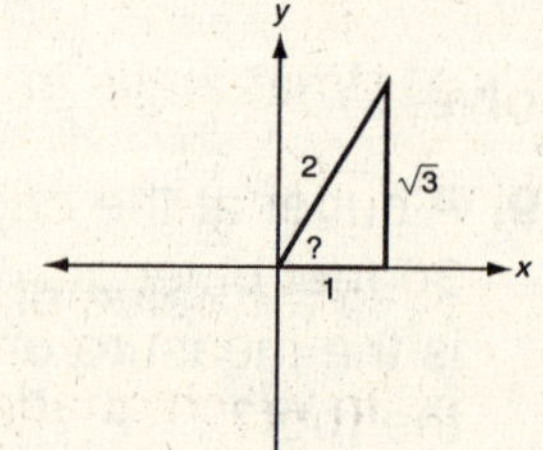

____

4. Add $(2\pi)n$ to each angle to find all possible values.

____

Holt Algebra 2

Reteach

LESSON 13-4

Inverses of Trigonometric Functions (continued)

When their domains are restricted, the inverse trigonometric relations can be defined as functions. These inverse trigonometric functions can be distinguished from the inverse trigonometric relations by the capital letters used to denote the functions.

Use the range of the inverse functions to help you evaluate the functions.

- For $\mathrm{Sin}^{-1}a = \theta$, θ lies in Quadrants I and IV and $-\dfrac{\pi}{2} \le \theta \le \dfrac{\pi}{2}$.
- For $\mathrm{Cos}^{-1}a = \theta$, θ lies in Quadrants I and II and $0 \le \theta \le \pi$.
- For $\mathrm{Tan}^{-1}a = \theta$, θ lies in Quadrants I and IV and $-\dfrac{\pi}{2} < \theta < \dfrac{\pi}{2}$.

Note that for $\mathrm{Sin}^{-1}a$ and $\mathrm{Cos}^{-1}a$, the terminal side of θ can lie on one of the axes.

For example: $\mathrm{Sin}^{-1}(-1) = -\dfrac{\pi}{2}$ or $-90°$.

Evaluate $\mathrm{Tan}^{-1}\left(-\dfrac{\sqrt{3}}{3}\right)$.

Step 1 Decide which quadrant θ lies in.
The value of the tangent is negative.
Tangent is negative in Quadrant IV.

Step 2 Evaluate the function.

$$\mathrm{Tan}\left(-\dfrac{\pi}{6}\right) = \left(-\dfrac{\sqrt{3}}{3}\right)$$

so $\mathrm{Tan}^{-1}\left(-\dfrac{\sqrt{3}}{3}\right) = -\dfrac{\pi}{6}$

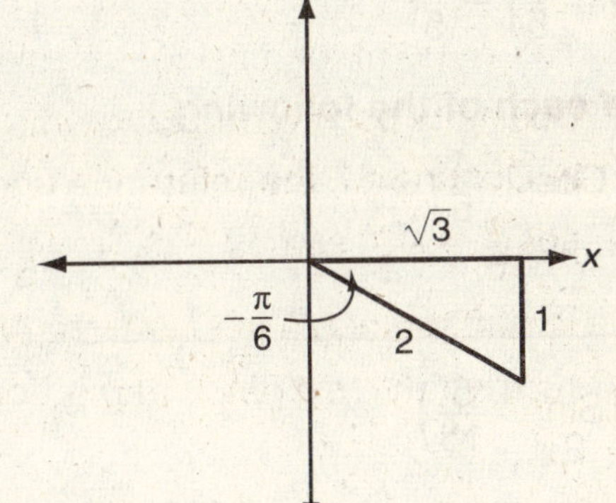

Complete to evaluate each trigonometric function.

5. $\mathrm{Cos}^{-1}\dfrac{\sqrt{2}}{2}$

   a. Is the value of cosine negative or positive? ________________

   b. In which quadrant does θ lie? ________________

   c. What angle in the range has cosine of $\dfrac{\sqrt{2}}{2}$? ________________

6. $\mathrm{Sin}^{-1}\left(-\dfrac{1}{2}\right)$

   a. Is the value of sine negative or positive? ________________

   b. In which quadrant does θ lie? ________________

   c. What angle in the range has sine of $-\dfrac{1}{2}$? ________________

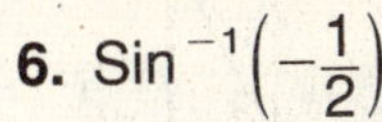

Holt Algebra 2

Challenge

 Combination Trigonometric Functions

When evaluating an inverse trigonometric function, it is helpful to think of the result as an angle. For example, $\text{Sin}^{-1}x$ can be thought of as an angle whose sine is x. The other common notation for $\text{Sin}^{-1}x$ is arcsine x and is read as an angle whose sine is x.

Thinking of the inverse trigonometric functions in this manner will help you evaluate combinations of trigonometric functions and inverse trigonometric functions. For example, consider this expression.

$$\sin\left(\text{Cos}^{-1}\frac{3}{5}\right)$$

The expression directs you to find the sine of an angle θ such that $\theta = \text{Cos}^{-1}\frac{3}{5}$. This is interpreted as θ is an angle whose cosine is $\frac{3}{5}$. Think of θ as an angle in a right triangle as shown at right. The cosine of angle θ is $\frac{3}{5}$ since that is the ratio of the adjacent side to the hypotenuse. The third side of the triangle is 4, and the ratio of the opposite side to the hypotenuse is $\frac{4}{5}$. It follows that $\sin\left(\text{Cos}^{-1}\frac{3}{5}\right) = \frac{4}{5}$.

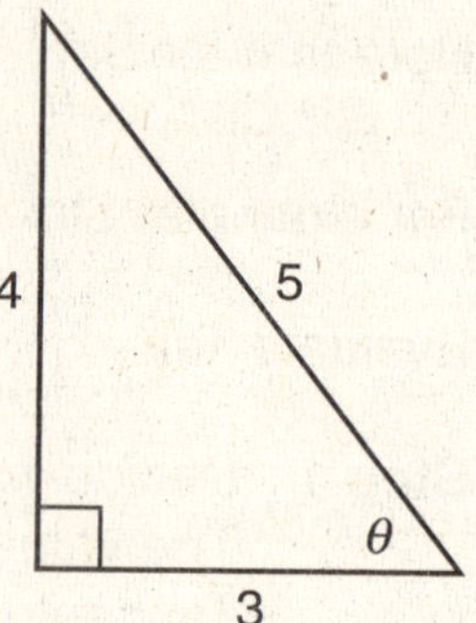

Evaluate each of the following.

1. $\sin\left(\text{Cos}^{-1}\frac{4}{5}\right)$

2. $\cos\left(\text{Tan}^{-1}\frac{3}{4}\right)$

____

3. $\tan\left(\text{Sin}^{-1}\frac{5}{13}\right)$

4. $\cos\left(\text{Sin}^{-1}\frac{12}{13}\right)$

____

5. $\sec\left(\text{Tan}^{-1}\frac{7}{24}\right)$

6. $\csc\left(\text{Sin}^{-1}\frac{9}{41}\right)$

____

7. $\cos\left(\text{Sin}^{-1}x\right)$ where $-1 \le x \le 1$
   (*Hint:* Think of x as a ratio and use the Pythagorean Theorem.)

____

8. $\csc\left(\text{Tan}^{-1}\frac{1}{x}\right)$ where $x > 0$
   (*Hint:* Draw a triangle and use the Pythagorean Theorem.)

____

Holt Algebra 2

Problem Solving
Inverses of Trigonometric Functions

Rafe is concerned that some recently constructed buildings in his town do not comply with code restrictions. New buildings are limited to a maximum of 40 feet in height.

1. When working with angles of elevation from his eye level, Rafe realizes that he must allow for his own height, 5 feet 9 inches, in his calculations. Explain how he can do this.

____

2. On the building plans, the height of the new bank is 33 feet. Rafe calculates what the angle of elevation should be from 100 feet away if the bank is 33 feet tall.

   a. Label the diagram to show the height of the bank that Rafe will use in his calculations Mark the angle of elevation.

   b. Write a trigonometric function for the assumed angle of elevation, θ

   ____

   c. To the nearest tenth of a degree, what is the assumed angle of elevation?

   ____

   d. Using a clinometer, Rafe measures the angle of elevation to be 14.2°. Does the bank comply with the building code? Explain.

____

3. On the plans, the height of the new inn is 39 feet. Rafe finds the angle of elevation and compares it to what he measures.

   a. Predict the angle of elevation of the highest point on this building from a distance of 100 feet and allowing for Rafe's height. ____________

   b. Predict the angle of elevation of the highest point on a building that is 40 feet tall, allowing for Rafe's height. ____________

   c. If Rafe measures an angle of elevation of 19.0°, how does the height of the inn compare to its declared height of 39 feet? Explain.

   ____

   ____

4. Carrie says the angle of elevation of the top of the flagpole at school is 55.9° from a distance of 20 feet away and allowing for her height of 5 feet 6 inches.

   a. Write and evaluate an expression for the height of the flagpole to the nearest tenth of a foot. ____________

   b. What should be the angle of elevation if Rafe measures it from a distance of 50 feet away? Write and evaluate an expression.

____

Holt Algebra 2

LESSON 13-4 Reading Strategy
Follow a Procedure

You can find the inverse of trigonometric functions, such as sine and cosine, using a two-step operation.

	EXAMPLE	STEP 1	STEP 2
For the inverse sine function, use the sine inverse relation $\sin^{-1} a = \theta$.	Find all possible values of $\sin^{-1} \frac{1}{2}$.	Using the unit circle, find the values between 0 and 2π for which $\sin\theta = \frac{1}{2}$. $\sin\frac{\pi}{6} = \frac{1}{2}$, $\sin\frac{5\pi}{6} = \frac{1}{2}$	Find angles that are coterminal with the angles $\frac{\pi}{6}$ and $\frac{5\pi}{6}$. $\frac{\pi}{6} + (2\pi)n$, $\frac{5\pi}{6} + (2\pi)n$
For the inverse cosine function, use the cosine inverse relation $\cos^{-1} a = \theta$.	Find all possible values of $\cos^{-1} \frac{1}{2}$.	Using the unit circle, find the values between 0 and 2π for which $\cos\theta = \frac{1}{2}$. $\cos\frac{\pi}{3} = \frac{1}{2}$, $\cos\frac{5\pi}{3} = \frac{1}{2}$	Find angles that are coterminal with the angles $\frac{\pi}{3}$ and $\frac{5\pi}{3}$. $\frac{\pi}{3} + (2\pi)n$, $\frac{5\pi}{3} + (2\pi)n$

Answer each question.

1. Explain the meaning of the inverse trigonometric functions.

   ____

2. Find all values of θ between 0 and 2π for which $\sin\theta = \frac{\sqrt{3}}{2}$.

   ____

3. Find all values of θ between 0 and 2π for which $\sin\theta = \frac{\sqrt{2}}{2}$.

   ____

4. Find all values of θ between 0 and 2π for which $\cos\theta = \frac{\sqrt{3}}{2}$.

   ____

5. Find all possible values of $\cos^{-1} \frac{\sqrt{2}}{2}$.

   a. Find the values of θ between 0 and 2π for which $\cos\theta = \frac{\sqrt{2}}{2}$.

   ____

   b. Find angles that are coterminal with these angle values.

   ____

6. Find all possible values of $\sin^{-1} = 0$.

   a. Find the values of θ between 0 and 2π for which $\sin\theta = 0$.

   ____

   b. Find angles that are coterminal with these angle values.

   ____

   c. Express your answer to part b in degrees.

   ____

Holt Algebra 2

Practice A
The Law of Sines

Find the area of each triangle. Round to the nearest tenth.

1. 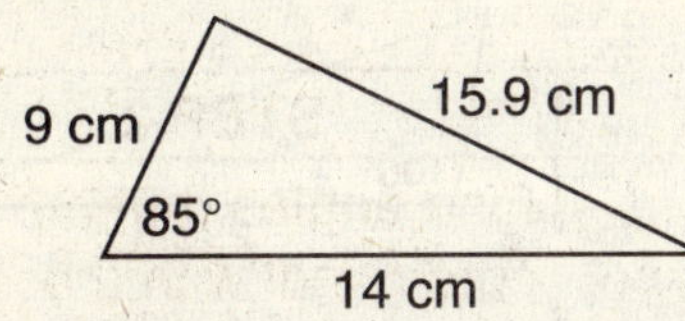

a. Write the formula for the area of a triangle.

____

b. Substitute the known values and evaluate.

____

2. 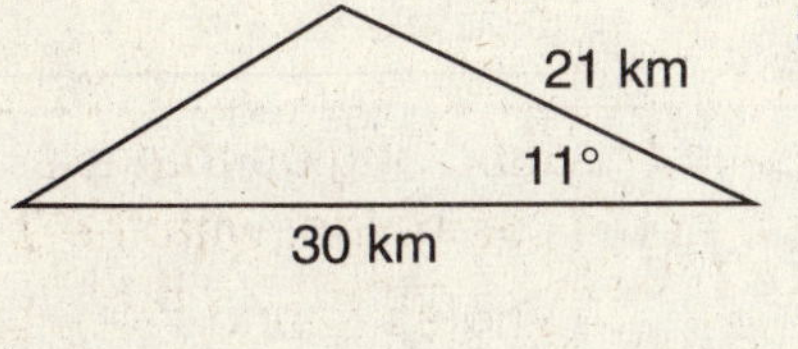

3. 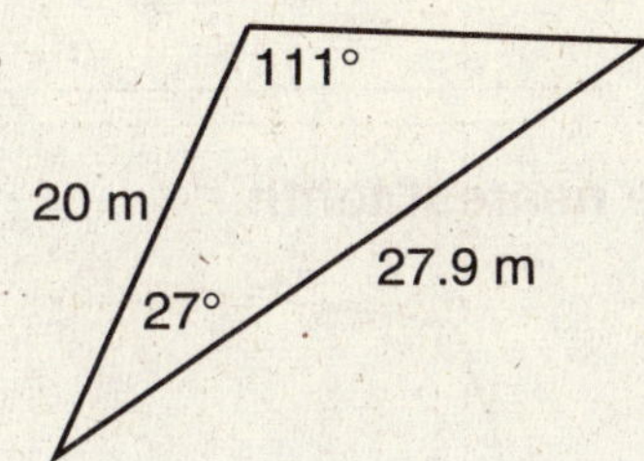

4. 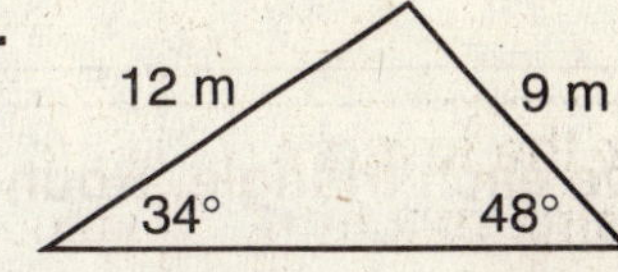

___________________   ___________________   ___________________

Solve each triangle. Round to the nearest tenth.

5. 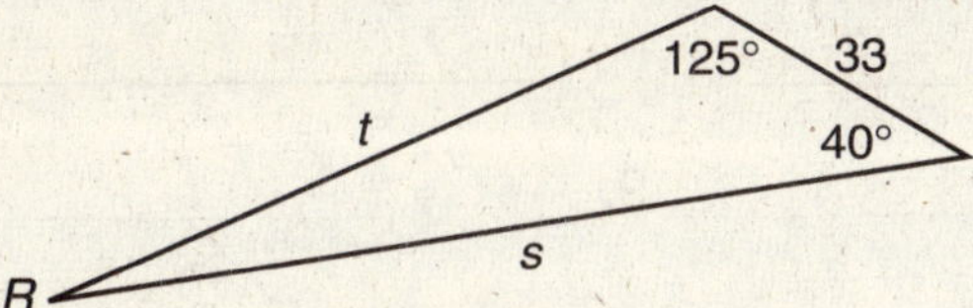

a. Find the measure of the third angle.

____

b. Use the Law of Sines to find the unknown side lengths.

____

6. 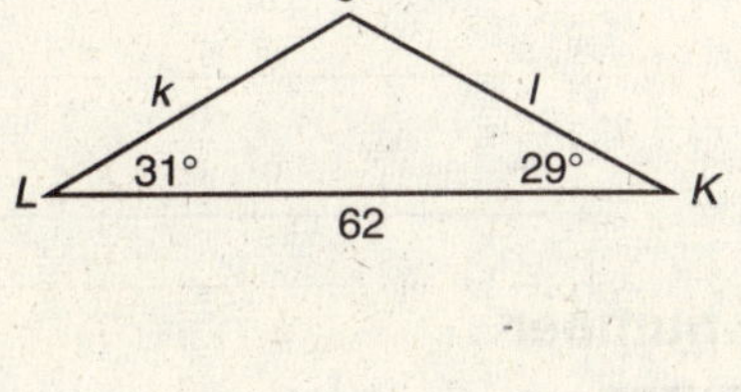

7. 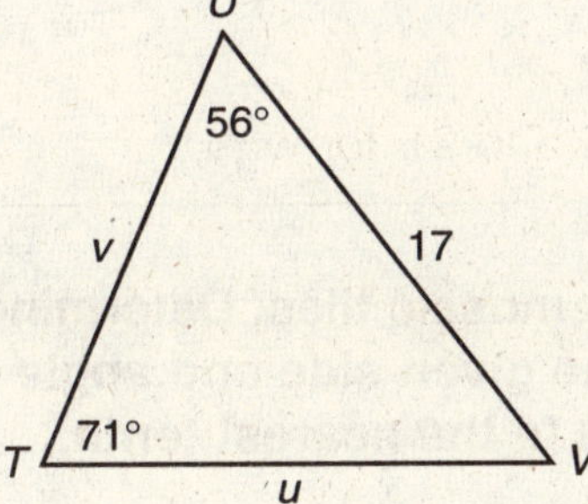

8. 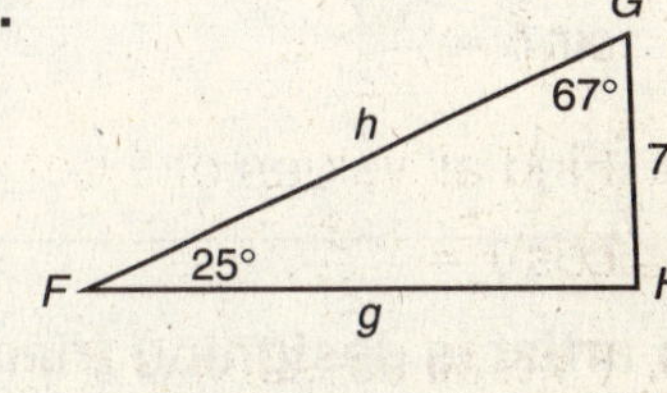

___________________   ___________________   ___________________

___________________   ___________________   ___________________

Solve.

9. Two sides of a triangular garden have 6 feet and 9 feet of edging.
To the nearest tenth, what is the area of the garden if
the angle formed by the edged sides is 80°? ___________________

Holt Algebra 2

LESSON 13-5

Practice B
The Law of Sines

Find the area of each triangle. Round to the nearest tenth.

1. 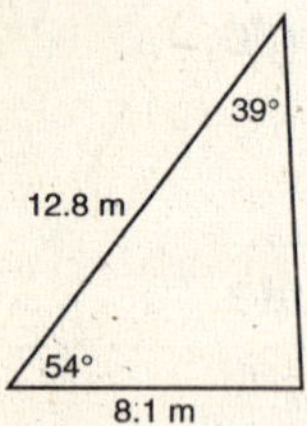

2. 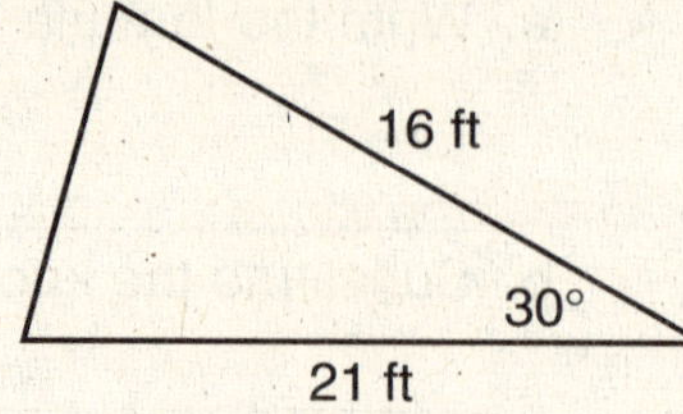

3. 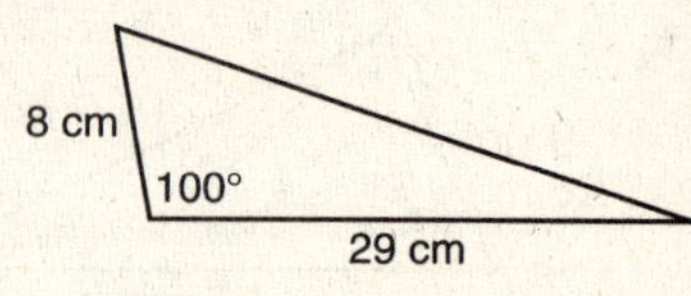

_______________ _______________ _______________

Solve each triangle. Round to the nearest tenth.

4. 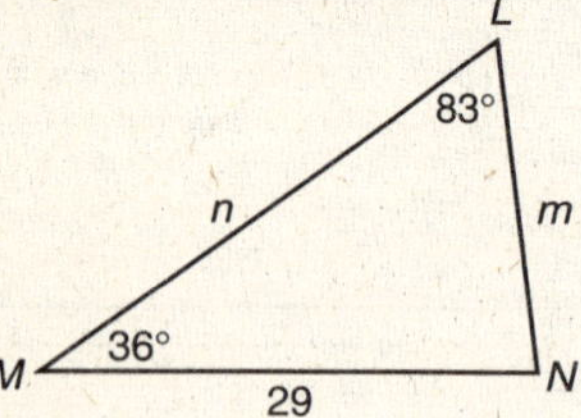

5. 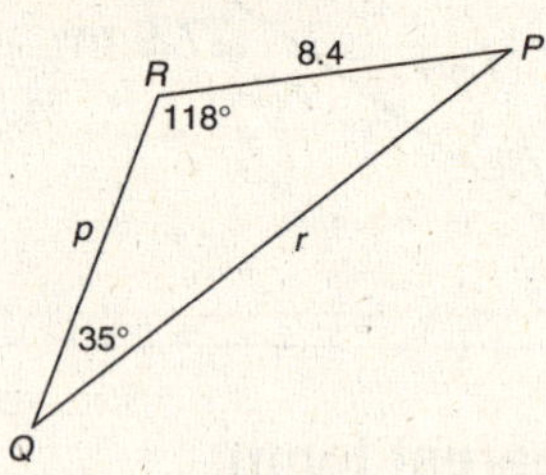

6. 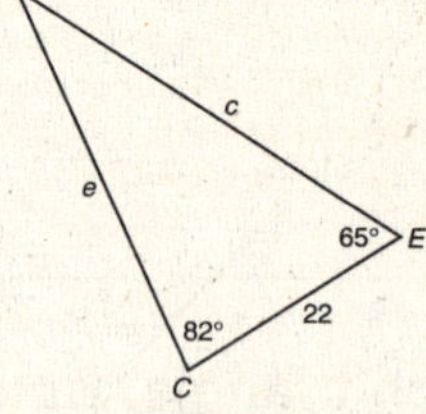

_______________ _______________ _______________

7. 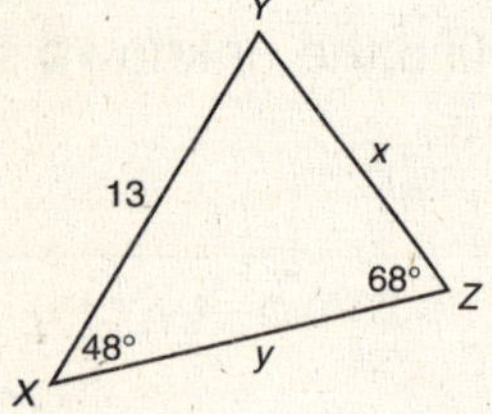

8. 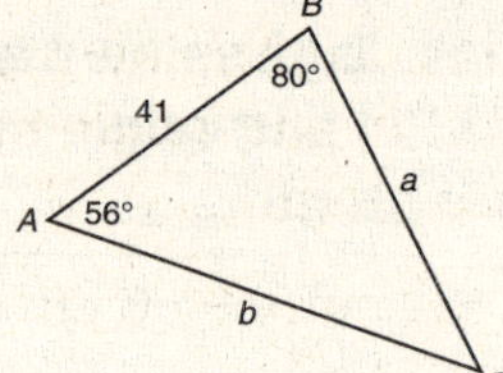

9. 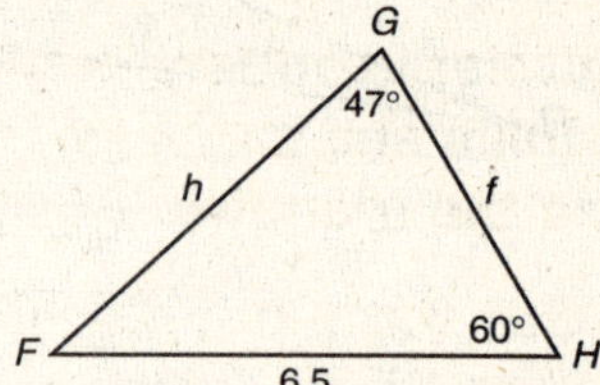

_______________ _______________ _______________

An artist is designing triangular mosaic tiles. Determine the number of triangles he can form from the given side and angle measures. Then solve the triangles. Round to the nearest tenth.

10. $a = 8$ cm, $b = 10$ cm, $A = 60°$

11. $a = 18$ cm, $b = 15$ cm, $A = 85°$

12. $a = 22$ cm, $b = 15$ cm, $A = 120°$

_______________ _______________ _______________

Solve.

13. Ann is creating a triangular frame. Two angles and the included side of the frame measure 64°, 58°, and 38 centimeters, respectively. What are the lengths of the other two sides of the frame to the nearest tenth of a centimeter?

____

Holt Algebra 2

LESSON
13-5
Practice C
The Law of Sines

Find the area of each triangle. Round to the nearest tenth.

1. 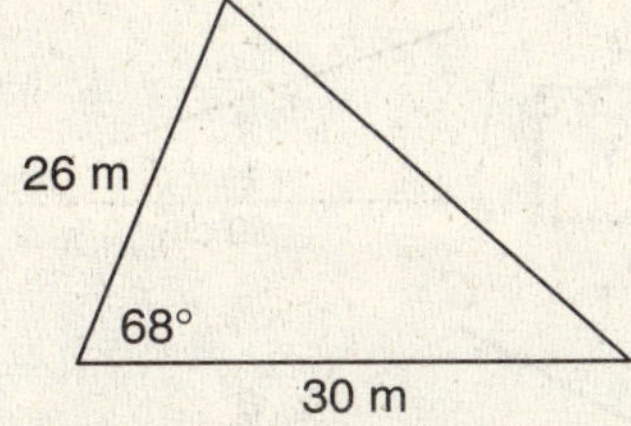

____

2. 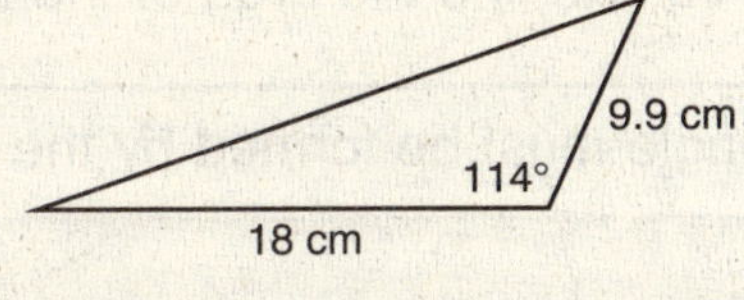

____

3. 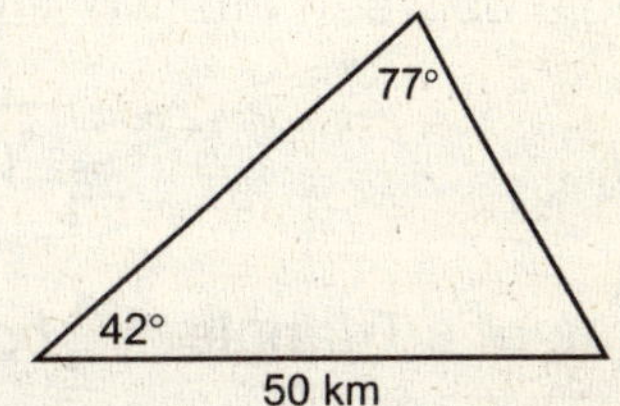

____

Solve each triangle. Round to the nearest tenth.

4. 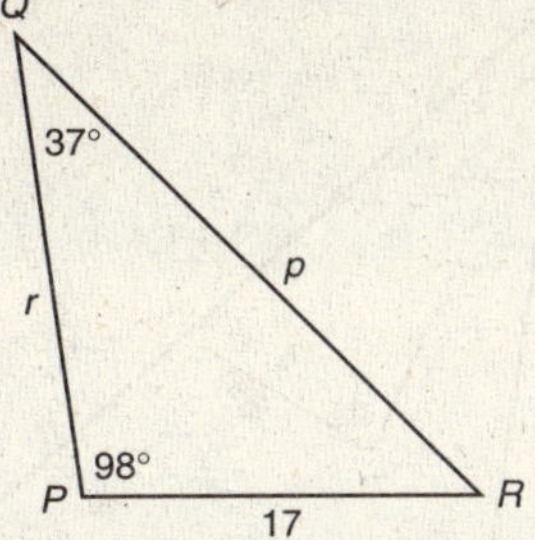

____

5. 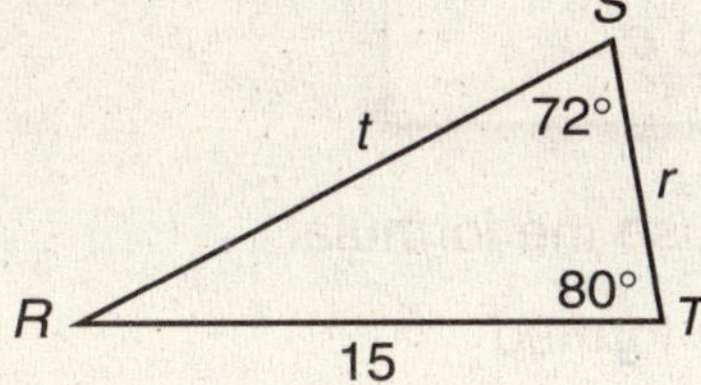

____

6. 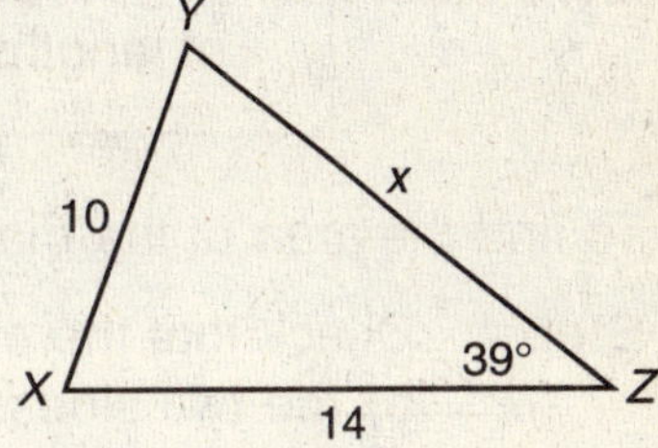

____

A jeweler is cutting stones into triangles. Determine the number of triangles she can form from the given side and angle measures. Then solve the triangles. Round to the nearest tenth.

7. $a = 8$ cm, $b = 42$ cm, $A = 12°$

8. $a = 21$ cm, $b = 6$ cm, $A = 90°$

9. $a = 7$ cm, $b = 7$ cm, $A = 90°$

10. $a = 27$ cm, $b = 30$ cm, $A = 55°$

11. $a = 29$ cm, $b = 22$ cm, $A = 75°$

12. $a = 4.6$ cm, $b = 9.2$ cm, $A = 30°$

Solve.

13. Margaret has two lengths of fence, 20 meters and 24 meters, for two sides of a triangular chicken pen. The third side will be on the north side of a barn. One fence length forms a 75° angle with the barn. How many different pens can she build if one fence is attached at the corner of the barn? What are all the possible lengths for the barn side of the pen?

____

Holt Algebra 2

Reteach
LESSON 13-5

The Law of Sines

If you know the lengths of two sides of a triangle and the measure of the angle between the two sides, you can find the area of the triangle.

> The angle must be formed by the two sides.

Area of a Triangle $= \frac{1}{2}bc\sin A$

> A is the measure of the angle between the sides of lengths b and c.

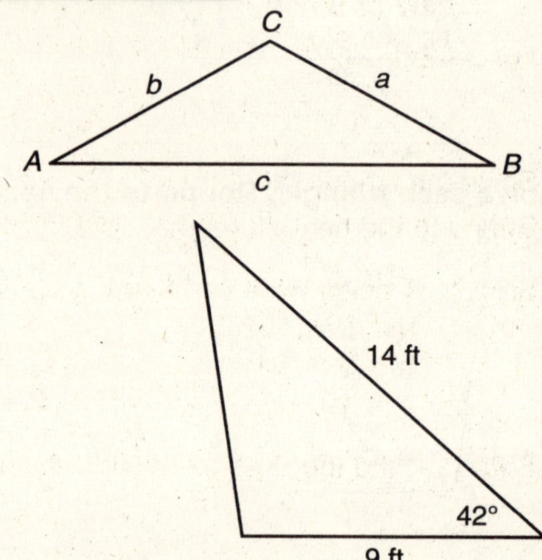

To find the area of the triangle, use the formula.

Step 1 Check that the angle is formed by the two sides.
Write the formula.

$$\text{Area} = \frac{1}{2}bc\sin A$$

Step 2 Substitute the known values into the formula.
$A = 42°$, $b = 9$, $c = 14$

$$\text{Area} = \frac{1}{2}(9)(14)\sin 42°$$

> Check that your calculator is set to degrees to compute the sine.

Step 3 Use a calculator to evaluate the area.
Area ≈ 42.1552282

Step 4 Round the answer to the nearest tenth.
Record the units.
The area is about 42.2 ft^2.

Complete to find the area of each triangle. Round to the nearest tenth.

1.

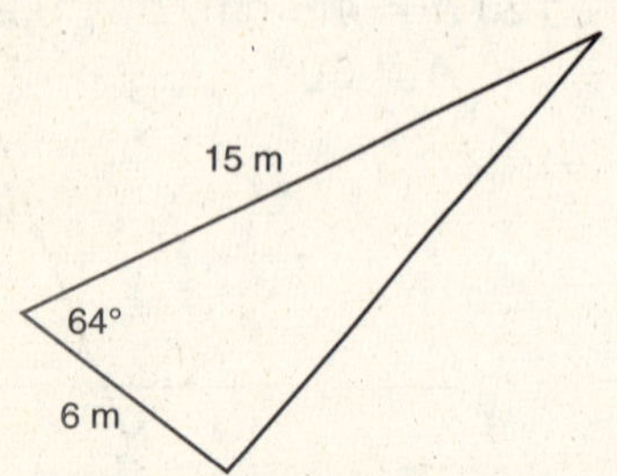

$$\text{Area} = \frac{1}{2}bc\sin A$$

$A = $______ $b = $______ $c = $______

____

2.

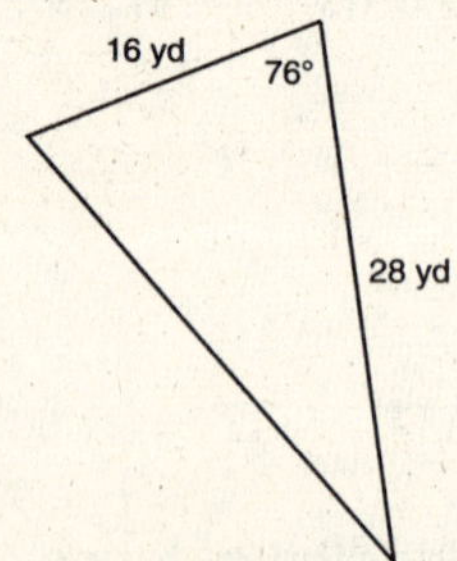

$$\text{Area} = \frac{1}{2}bc\sin A$$

$A = $______ $b = $______ $c = $______

____

Holt Algebra 2

<table><tr><td>**LESSON**
13-5</td><td></td></tr></table>

Reteach
The Law of Sines (continued)

Use the **Law of Sines** to solve a triangle if you know two angle measures and any side length **or** two side lengths and an angle measure that is not between them.

> "To solve a triangle" means to find the missing side lengths and angle measures.

Law of Sines

For $\triangle ABC$: $\dfrac{\sin A}{a} = \dfrac{\sin B}{b} = \dfrac{\sin C}{c}$

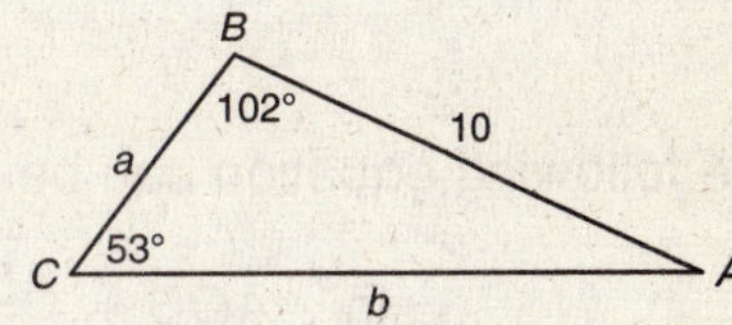

Solve the triangle at right.
Round to the nearest tenth.

Step 1 Decide what you know.
$m\angle B = 102°$
$m\angle C = 53°$
$c = 10$

Step 2 Find the missing angle measure.
$53° + 102° + m\angle A = 180°$
$m\angle A = 25°$

> The sum of the measures of the angles of a triangle is 180°.

Step 3 Use the Law of Sines to find the unknown side lengths.

$\dfrac{\sin A}{a} = \dfrac{\sin C}{c}$ $\qquad$ $\dfrac{\sin B}{b} = \dfrac{\sin C}{c}$

$\dfrac{\sin 25°}{a} = \dfrac{\sin 53°}{10}$ $\qquad$ $\dfrac{\sin 102°}{b} = \dfrac{\sin 53°}{10}$

> Use $\dfrac{\sin C}{c}$ since you know the measures of C and c.

$a\sin 53° = 10\sin 25°$ $\qquad$ $b\sin 53° = 10\sin 102°$

$a = \dfrac{10\sin 25°}{\sin 53°}$ $\qquad$ $b = \dfrac{10\sin 102°}{\sin 53°}$

$a \approx 5.3$ $\qquad\qquad$ $b \approx 12.2$

Complete to solve the triangle.

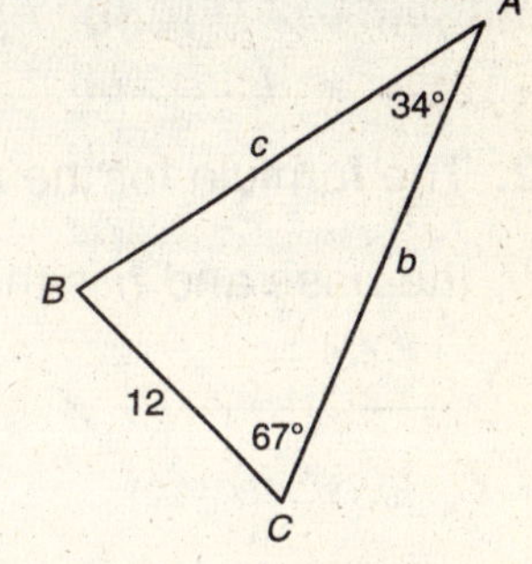

3. $m\angle A =$ __________ $m\angle C=$ __________ $a =$ __________

4. Find $m\angle B$.

____

5. Find b to the nearest tenth.

$\dfrac{\sin A}{a} = \dfrac{\sin B}{b}$, $\qquad$ $\dfrac{\sin 34°}{12} = \dfrac{\sin 79°}{b}$

____

6. Find c to the nearest tenth.

$\dfrac{\sin A}{a} = \dfrac{\sin C}{c}$

____

Holt Algebra 2

Challenge
LESSON 13-5 — *A Geometric Law of Sines*

The Law of Sines can be derived by a geometric argument. Consider the diagram below. The measure of $\angle ACB$ is $\frac{1}{2}$ the measure of $\angle AOB$. Right triangles AOD and BOD are congruent since they share a common leg (side OD) and each hypotenuse is a radius of the circle. So the $m\angle AOD = m\angle BOD = m\angle ACB$. Also $\overline{AD} = \overline{BD} = \frac{1}{2}c$.

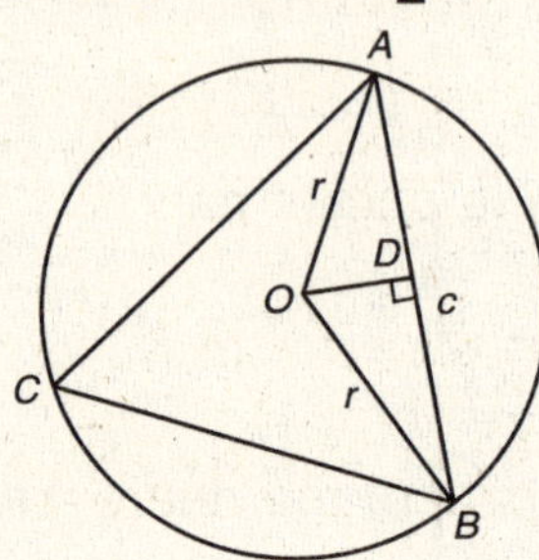

The following equation can be derived from the information given above.

$$\sin \angle AOD = \frac{AD}{AO} = \frac{\frac{c}{2}}{r} = \sin \angle ACB = \sin \angle C$$

This simplifies to $2r = \dfrac{c}{\sin \angle C}$, which tells us that that the ratio of the length of a side of a triangle to the sine of the opposite angle is equal to twice the radius of the circle circumscribed about the triangle. Thus we obtain the following, which leads immediately to the Law of Sines.

$$2r = \frac{a}{\sin \angle A} = \frac{b}{\sin \angle B} = \frac{c}{\sin \angle C}$$

Solve.

1. A buoy is anchored offshore to mark a sandbar. The straight shoreline at that location runs north and south. From two observation points on the shore 2.4 miles apart, the bearings to the buoy are S46°E and N22°E. How far is the buoy from the shore? What is the distance from the buoy to each of the observation points?

2. The formula for the area of a triangle is Area $= \frac{1}{2}bh$, where b is the length of the base and h is the height of the triangle. Show that Area $= \dfrac{a^2 \sin B \sin C}{2\sin A}$.

Holt Algebra 2

Problem Solving
The Law of Sines

In the middle of town, State and Elm streets meet at an angle of 40°. A triangular pocket park between the streets stretches 100 yards along State Street and 53.2 yards along Elm Street. Hoa and Cat walk around the pocket park every day at lunchtime.

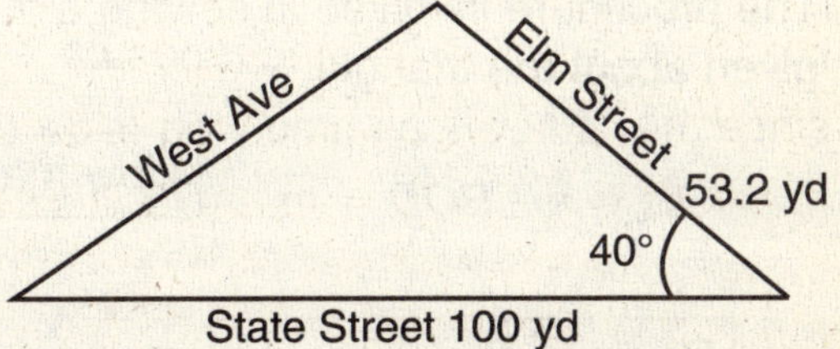

1. Hoa would like to know the area of the pocket park.

   a. Write a formula for the area of the pocket park using the given dimensions.

   ____

   b. What is the area of the pocket park to the nearest tenth of a square yard?

   ____

2. Cat determines that the total distance around the pocket park is 221.6 yards.

   a. Write and evaluate an expression for the length of the park along West Avenue.   ________________________

   b. Use the Law of Sines to find $\angle S$ to the nearest degree, the angle that West Avenue makes with State Street.   ________________________

   c. Write an expression for the angle that West Avenue makes with Elm Street.

   ____

3. West Avenue makes angles of 55° with Main Street and 80° with Third Street. The distance from Main to Third along West Avenue is 40 yards. Hoa contracts to design a pocket park for the acute-angled triangular area enclosed by these streets.

   a. What is the measure of the third angle of the triangular area?   ________________________

   b. Write and evaluate an expression for the distance from West Avenue to Third Street along Main Street.   ________________________

Choose the letter for the best answer.

4. Hoa wants to plant palm trees 8 feet apart along the side of the park on Third Street. Which expression gives the number of trees she will need?

   A $\dfrac{40 \sin 55°}{\sin 45°}$ **C** $\dfrac{5 \sin 45°}{\sin 55°}$

   B $\dfrac{5 \sin 55°}{\sin 45°}$ **D** $\dfrac{40 \sin 45°}{\sin 55°}$

5. What is the area of the new pocket park that Hoa is designing?

   F 913 yd^2

   G 1004 yd^2

   H 1207 yd^2

   J 1398 yd^2

Holt Algebra 2

LESSON 13-5 Reading Strategy
Use a Flowchart

The procedure involved in solving a triangle depends upon the information given about the triangle.

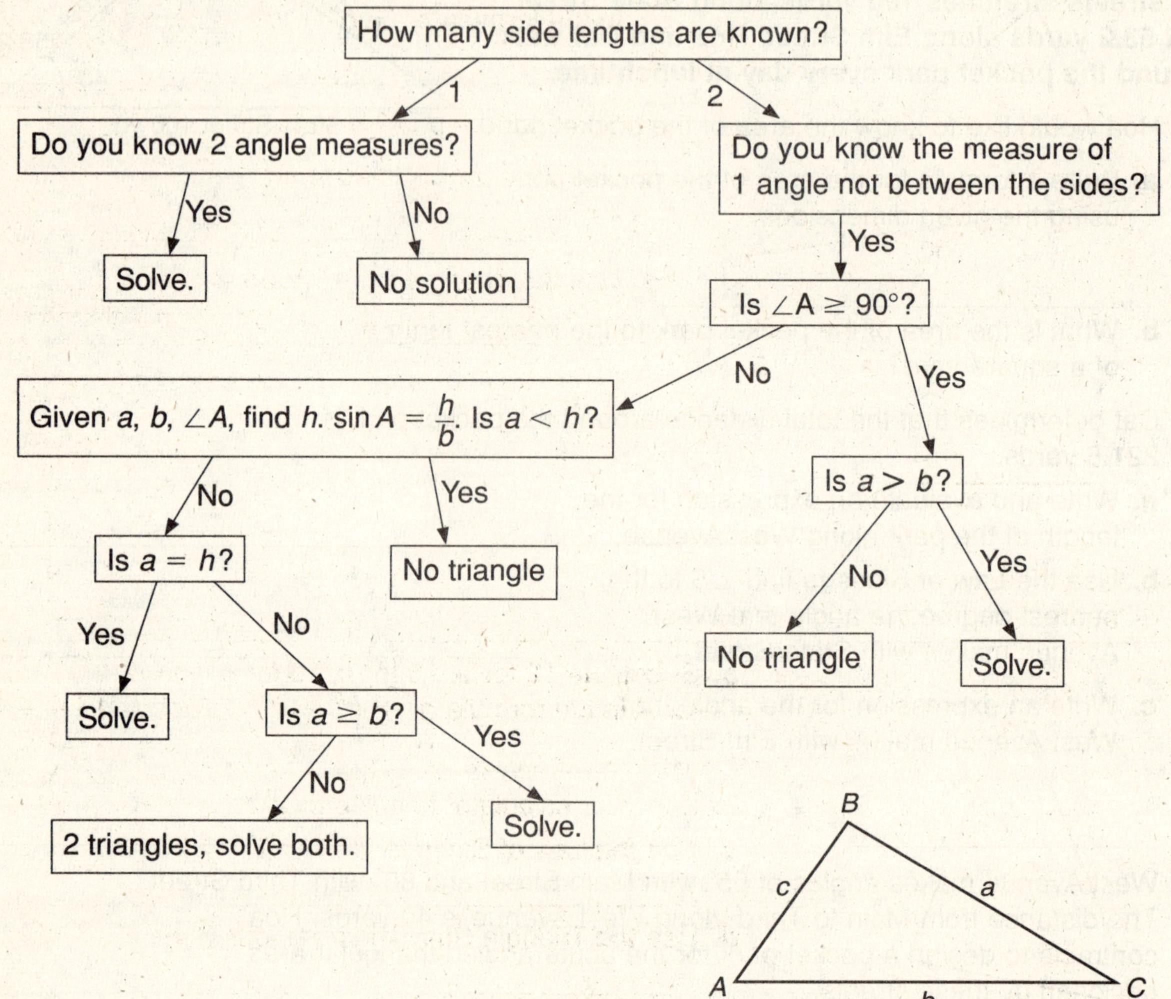

Trace a path through the flowchart for each possible triangle and state the outcome.

1. $b = 8$, m$\angle A = 55°$, m$\angle C = 45°$

2. $a = 13$, $b = 9$, m$\angle A = 80°$

3. $a = 4$, $b = 10$, m$\angle A = 95°$

4. $a = 8$, $b = 12$, m$\angle A = 35°$

5. $a = 15$, $b = 11$, m$\angle A = 102°$

6. $a = 5$, $b = 7$, m$\angle A = 62°$

Holt Algebra 2

Practice A
The Law of Cosines

Solve each triangle. Round to the nearest tenth.

1.

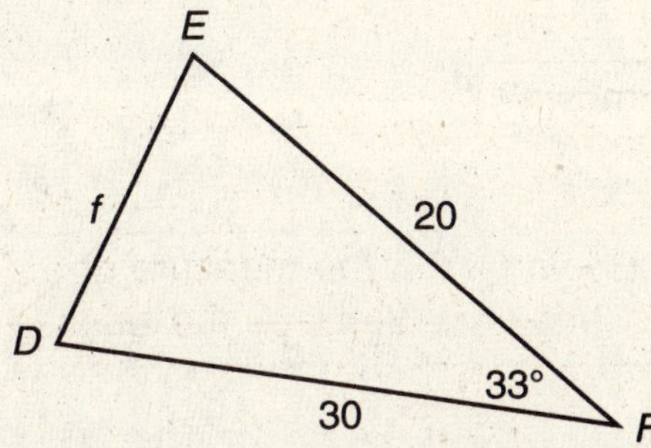

a. Substitute f for a, 20 for b, 30 for c, and 33° for A in the formula $a^2 = b^2 + c^2 - 2bc\cos A$.

____

b. Solve for the positive value of a. _______________

c. Use the Law of Sines to find $m\angle E$.

____

d. Use the Triangle Sum Theorem to find $m\angle D$.

____

2.

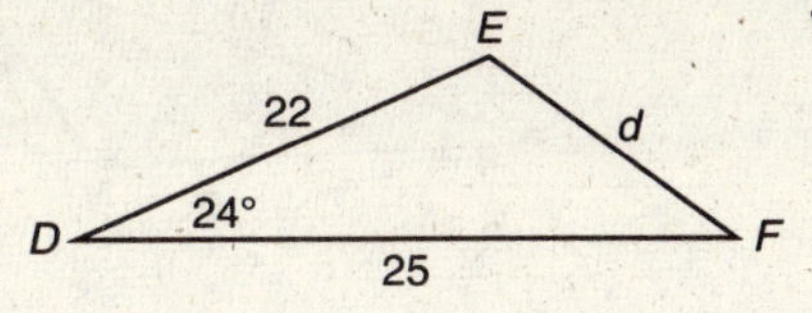

3.

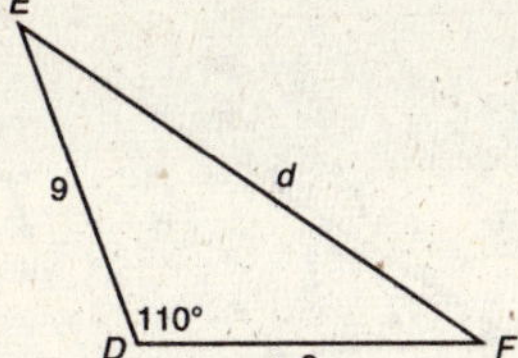

4.

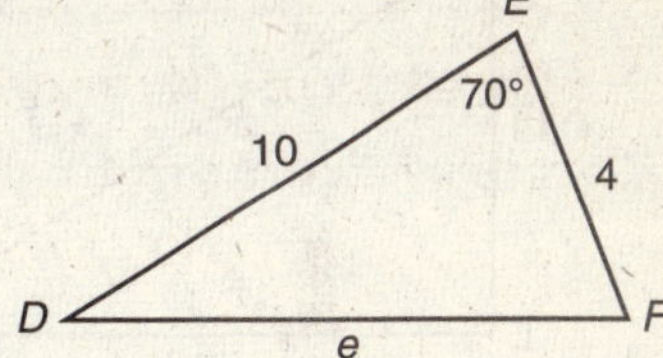

____

5.

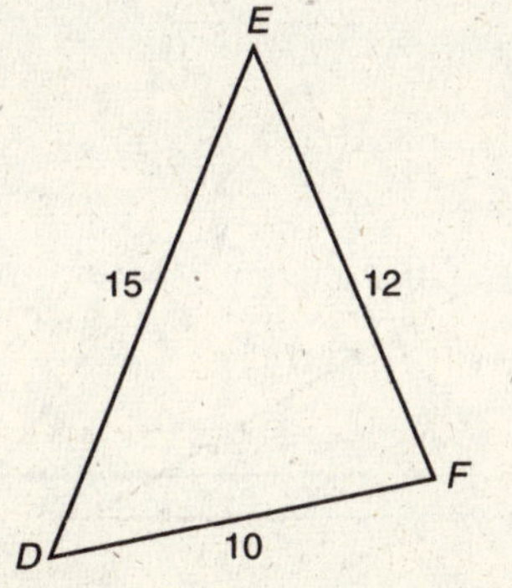

a. Substitute 12 for a, 15 for b, 10 for c, and D for A in the formula $a^2 = b^2 + c^2 - 2bc\cos A$.

____

b. Use your calculator to solve for $m\angle D$. _______________

c. Use the Law of Sines to find $m\angle E$.

____

d. Use the Triangle Sum Theorem to find $m\angle F$.

____

6.

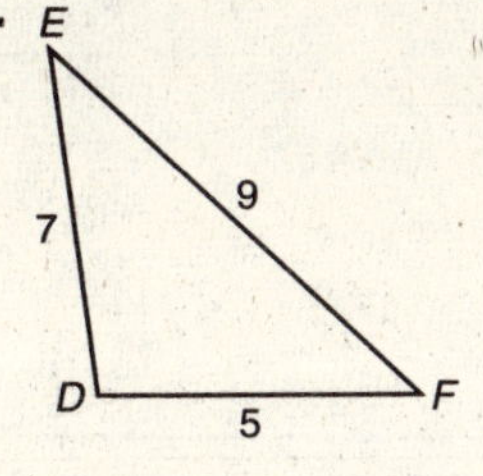

7.

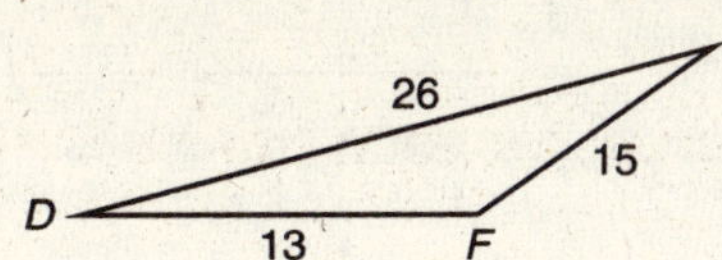

8.

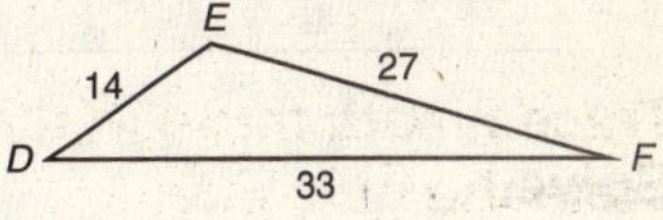

Find the area using Heron's Formula, Area $= \sqrt{s(s-a)(s-b)(s-c)}$.

9. A triangle with side lengths of 14 meters, 19 meters, and 21 meters

____

10. A triangle with side lengths 7.5 miles, 11 miles, and 13 miles

____

Holt Algebra 2

Name _________________________________ Date __________ Class __________

Practice B
The Law of Cosines

Use the given measurements to solve each triangle. Round to the nearest tenth.

1. 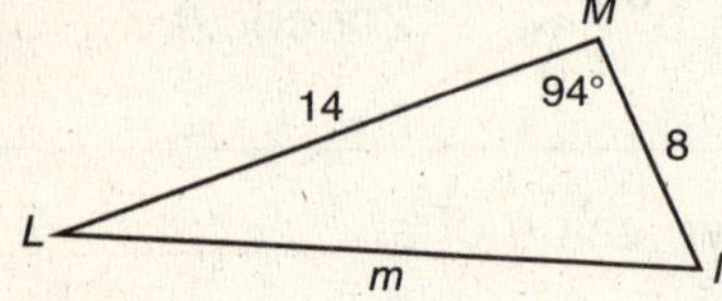

2. 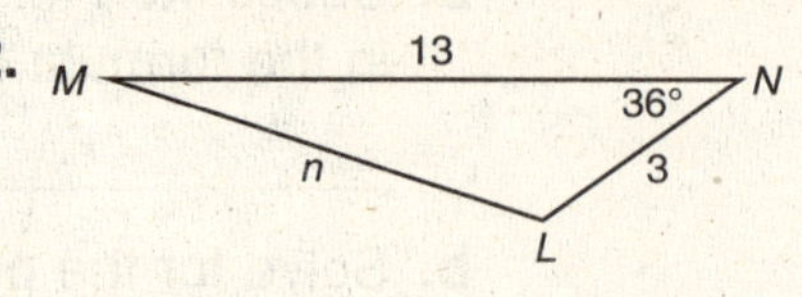

3. 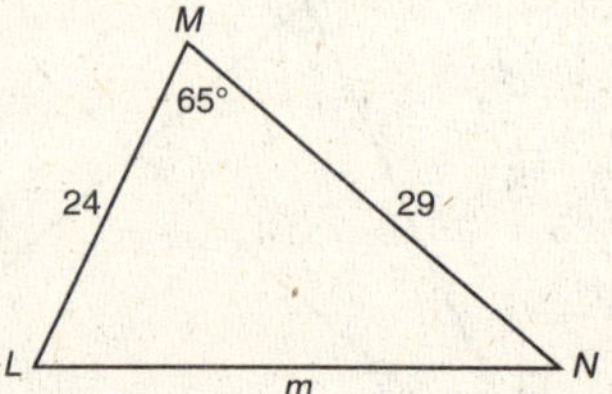

____

4. 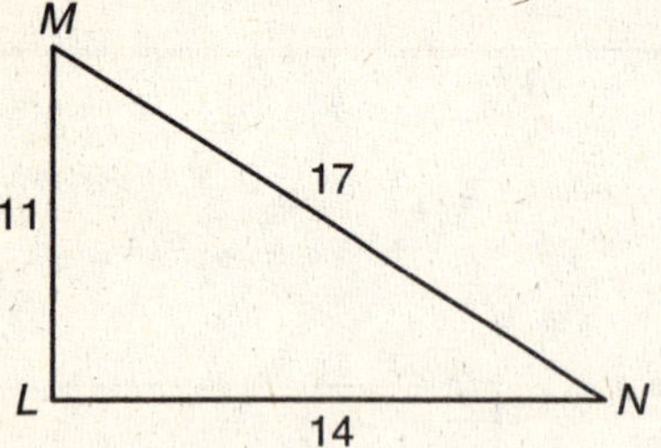

5. 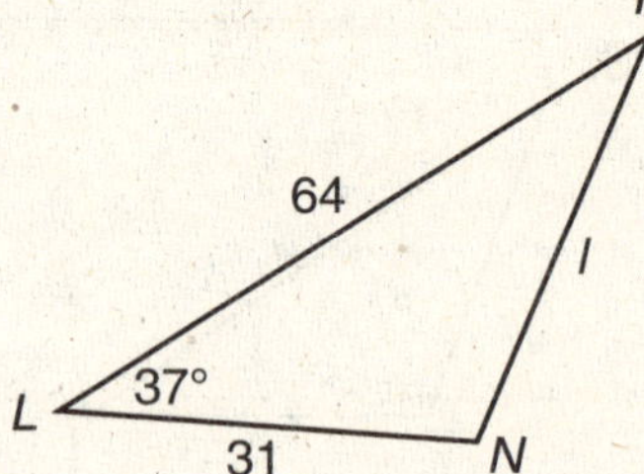

6. 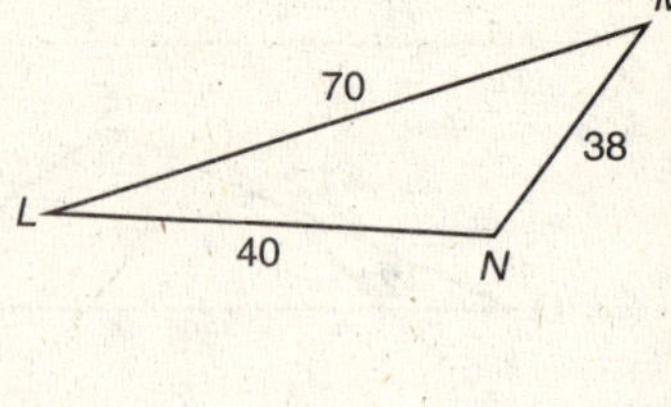

____

7. 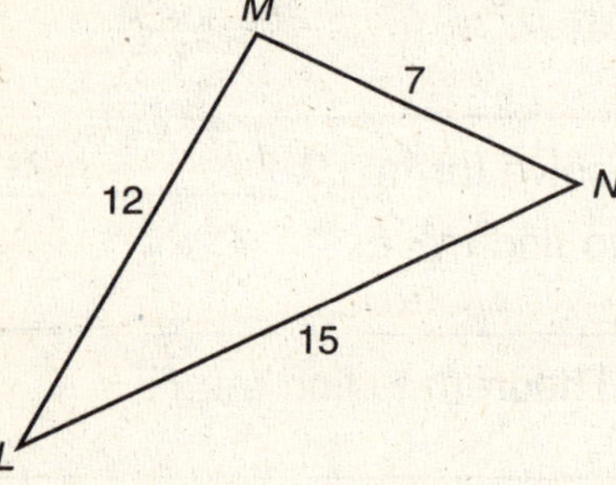

8. 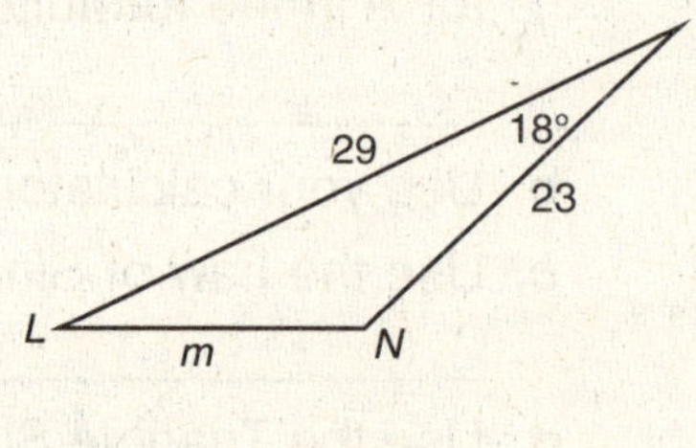

9. 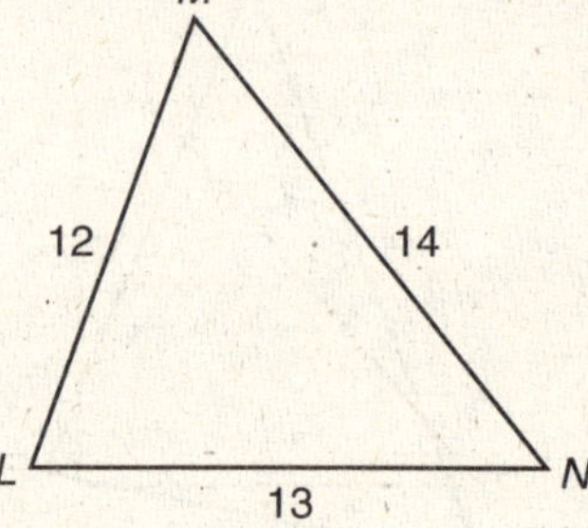

____

Solve.

10. A postal airplane leaves Island A and flies 91 miles to Island B. It drops off and picks up mail and flies 63 miles to Island C. After unloading and loading mail, the plane returns to Island A at an average rate of 300 miles per hour. How long does it take the pilot to travel from Island C to Island A?

____

11. A statue is erected on a triangular marble base. The lengths of the sides of the triangle are 12 feet, 16 feet, and 18 feet. What is the area of the region at the base of the statue?

____

Holt Algebra 2

Practice C
The Law of Cosines

Use the given measurements to solve each triangle. Round to the nearest tenth.

1.

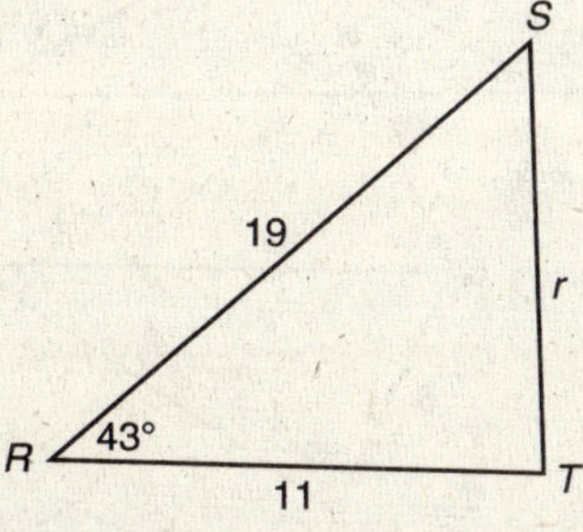

____

2.

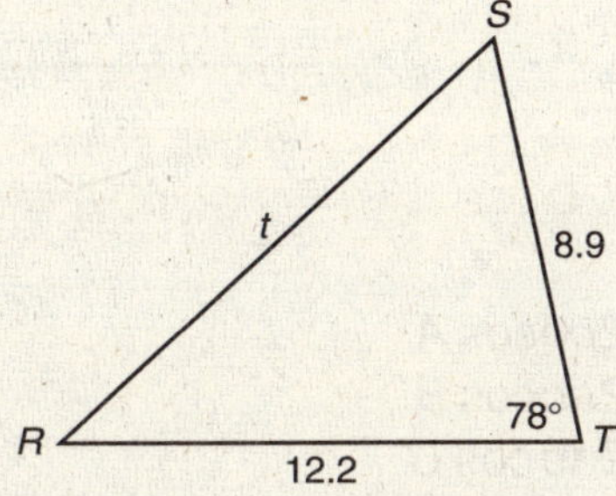

____

3.

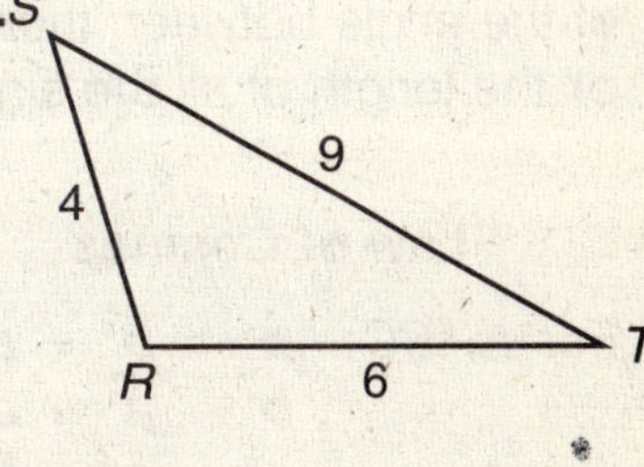

____

4.

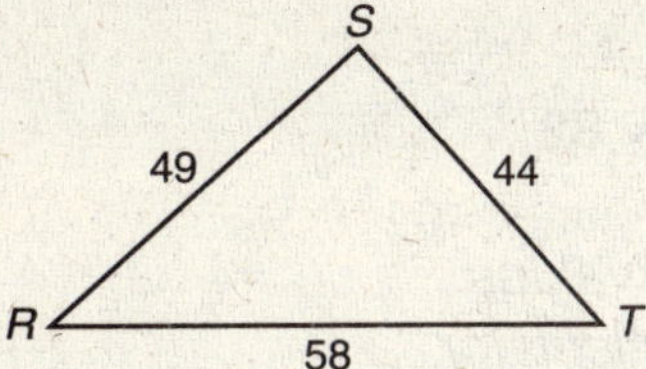

____

5.

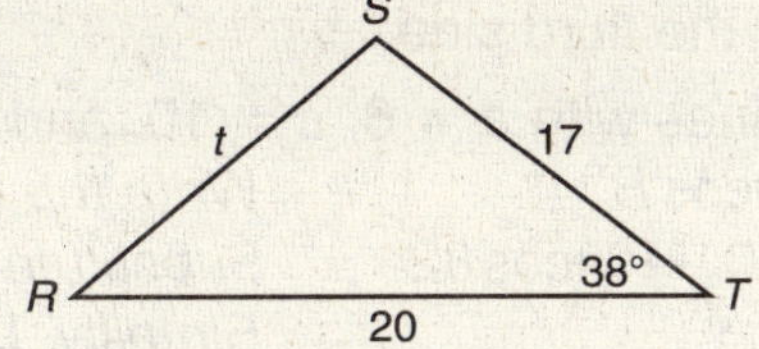

____

6.

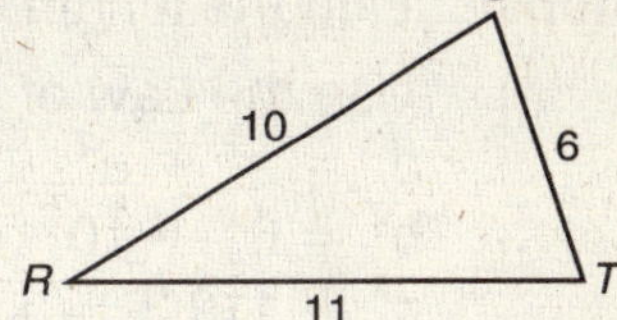

____

7.

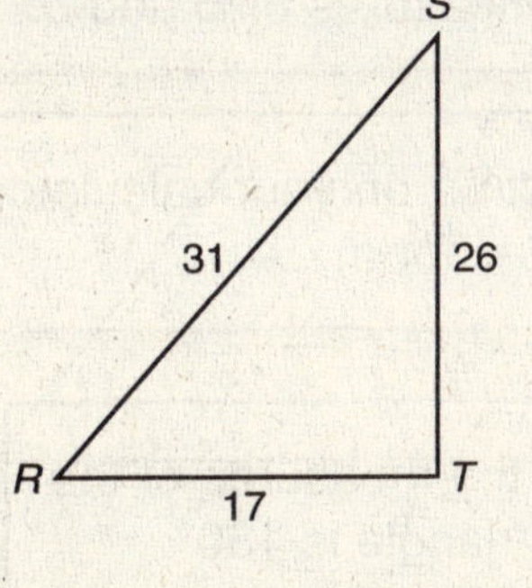

____

8.

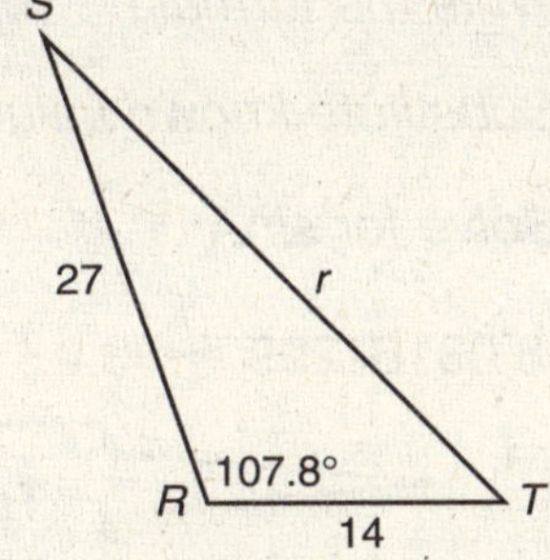

____

9.

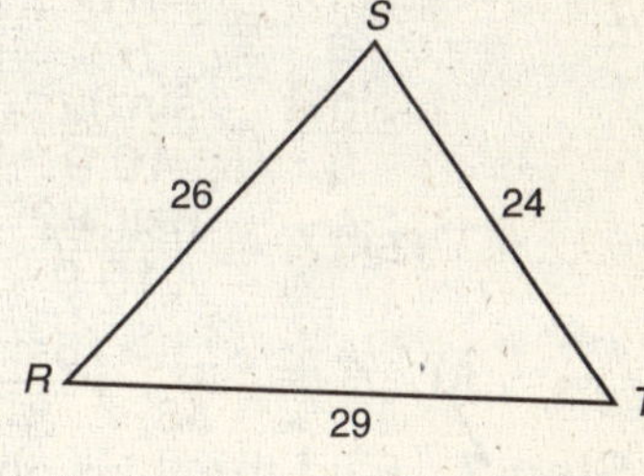

____

Solve.

10. The rotunda of a shopping mall is in the shape of a triangle. Members of a local scout troop plan to walk around the outer edge of the rotunda 5 times to increase awareness of physical fitness. If they walk at an average speed of 3 miles per hour, how long should it take them to the nearest minute?

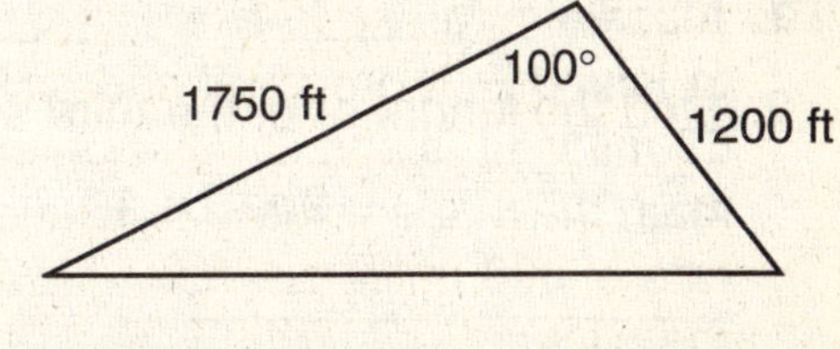

____

11. There is a fountain in a triangular garden in front of City Hall. The lengths of the sides of the triangle are 14 meters, 15 meters, and 21 meters. What is the area of the garden?

____

Holt Algebra 2

Reteach
13-6 *The Law of Cosines*

Use the **Law of Cosines** to solve a triangle
if you know the length of two sides and the measure
of the angle between them
or the length of all the sides.

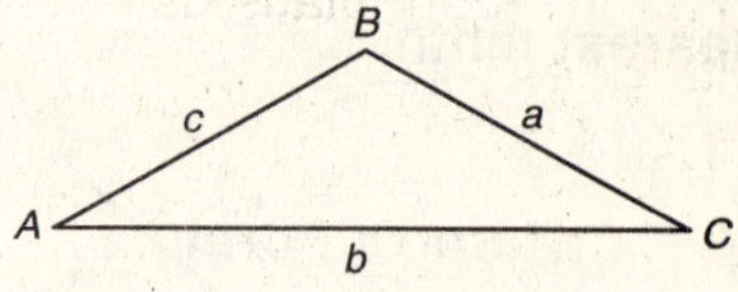

Law of Cosines

For $\triangle ABC$: $a^2 = b^2 + c^2 - 2bc\cos A$
$b^2 = a^2 + c^2 - 2ac\cos B$
$c^2 = a^2 + b^2 - 2ab\cos C$

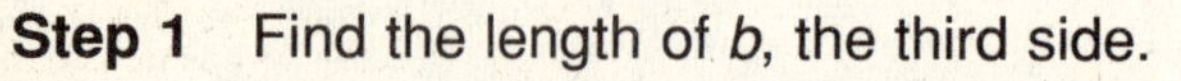

Solve the triangle at right.
Round to the nearest tenth.

Step 1 Find the length of b, the third side.

Use the Law of Cosines with $a = 6$, $c = 10$, and $m\angle B = 82°$.
$b^2 = a^2 + c^2 - 2ac\cos B$ *Write the formula.*
$b^2 = 6^2 + 10^2 - 2(6)(10)\cos 82°$ *Substitute known values.*
$b^2 \approx 119.3$ *Simplify. Use a calculator.*
$b \approx 10.9$ *Solve for the positive value of b.*

Step 2 Find an angle measure. Use the Law of Sines.

$\dfrac{\sin A}{a} = \dfrac{\sin B}{b}$ *Write the formula.*

$\dfrac{\sin A}{6} \approx \dfrac{\sin 82°}{10.9}$ *Substitute known values.*

$\sin A \approx \dfrac{6\sin 82°}{10.9}$ *Solve for sin A.*

Use $\dfrac{\sin B}{b}$ since you know the measures of B and b.

Use $\sin^{-1}$ on your calculator to solve for $m\angle A$.

$m\angle A \approx \text{Sin}^{-1}\left(\dfrac{6\sin 82°}{10.9}\right) \approx 33.03161299 \approx 33.0$

Step 3 Find the third angle measure.

$82° + 33° + m\angle C \approx 180°$

$m\angle C \approx 65°$

The sum of the measures of the angles of a triangle is 180°.

Complete to solve the triangle. Round to the nearest tenth.

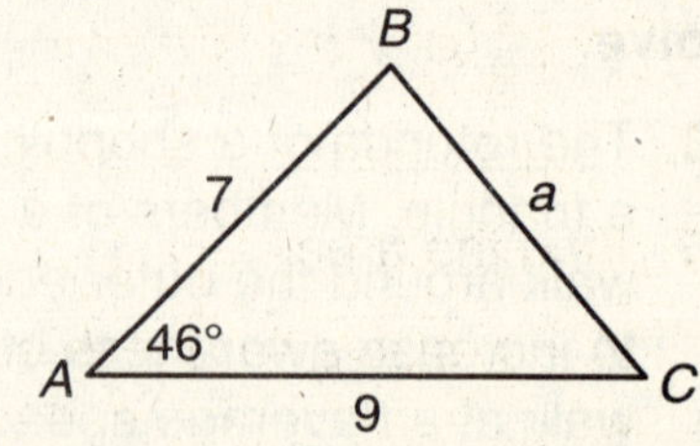

1. $m\angle A=$ _________ $b =$ _________ $c =$ _________

2. Find the length of a, the third side.

$a^2 = b^2 + c^2 - 2bc\cos A$

____

3. Find the measure of $\angle B$.

$\dfrac{\sin A}{a} = \dfrac{\sin B}{b}$

____

4. Find the measure of $\angle C$.

____

Holt Algebra 2

LESSON 13-6

Reteach

The Law of Cosines (continued)

Heron's Formula is used to find the area of a triangle given the lengths of its sides.

> Notice that *s* is half the perimeter of the triangle.

Heron's Formula

For $\triangle ABC$, with $s = \frac{1}{2}(a + b + c)$,

$$\text{Area} = \sqrt{s(s - a)(s - b)(s - c)}$$

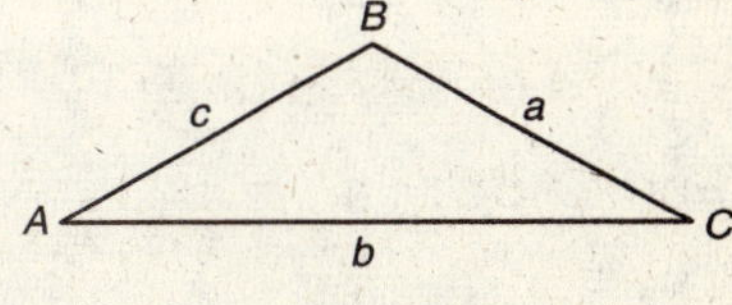

To find the area of the triangle at right, use Heron's Formula since the lengths of all the sides are known.

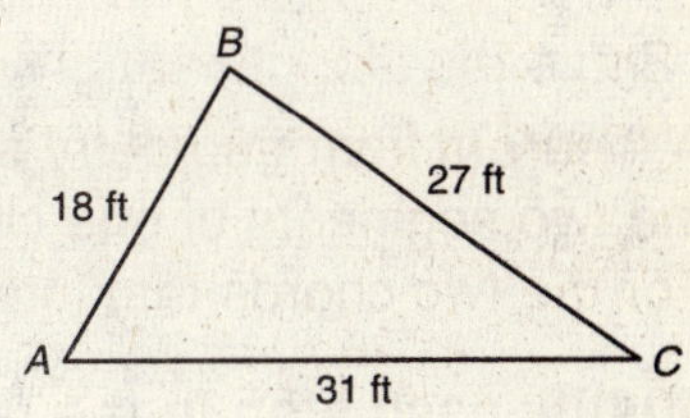

Step 1   Find the value of *s*.

$a = 27$, $b = 31$, and $c = 18$

$s = \frac{1}{2}(a + b + c)$ *Write the formula.*

$s = \frac{1}{2}(27 + 31 + 18)$ *Substitute known values.*

$s = 38$ *Simplify.*

Step 2   Find the area. Use Heron's Formula.
Round to the nearest foot.

$A = \sqrt{s(s - a)(s - b)(s - c)}$ *Write the formula.*

$A = \sqrt{38(38 - 27)(38 - 31)(38 - 18)}$ *Substitute. Use s = 38.*

$A = \sqrt{38(11)(7)(20)} = \sqrt{58{,}520}$ *Simplify.*

$A \approx 241.9090738$ *Find the square root.*

$A \approx 242 \text{ ft}^2$ *Round the answer.*

> Use a calculator.

Complete to find the area. Round to the nearest whole unit.

5. $a =$ _________ $b =$ _________ $c =$ _________

6. Find *s*.

$s = \frac{1}{2}(a + b + c)$

____

7. Find the area.

$A = \sqrt{s(s - a)(s - b)(s - c)}$

____

Holt Algebra 2

LESSON 13-6 Challenge
A Circular Derivation

The Law of Cosines can be derived by a geometric argument. Consider triangle *ABC* below. Vertex *B* is placed at the center of a circle, vertex *C* is located on the circle, and side *a* is the radius of the circle. Sides *AC* and *BC* are extended so that they intersect the circle at points *A'* and *B'*. Since *B'C* is a diameter of the circle, triangle *A'B'C* is a right triangle since $\angle A'$ intercepts a semicircle.

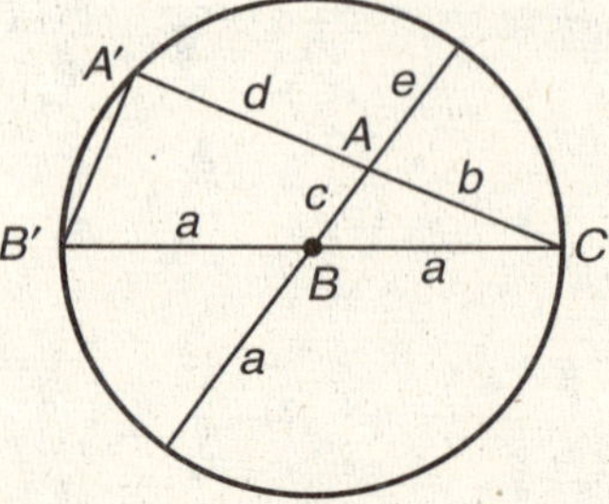

A theorem from geometry states that if two chords intersect inside a circle, then the product of the two segments of one chord is equal to the product of the two segments of the other chord. For the two chords that intersect at point *A*, we therefore have $bd = (a + c)e$.

Justify each step in the derivation of the Law of Cosines.

1. $bd = (a + c)(a - c)$ _______________________________

2. $b(2a\cos C - b) = a^2 - c^2$

   ____

3. $2ab\cos C - b^2 = a^2 - c^2$

   ____

4. $c^2 = a^2 + b^2 - 2ab\cos C$

   ____

5. A worker is installing an infrared security scanning device in the corner of a living room 6 inches below the ceiling as shown in the figure at right. She needs to set the scan angle on the scanning device so that it will scan from one corner of the room to the opposite corner of the room. The room measures 18 feet by 14 feet and is 9 feet high. What is the angle θ that the device should be set to scan?

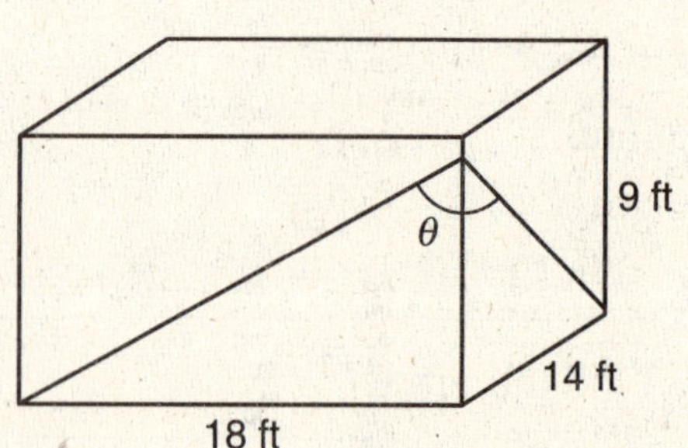

____

Holt Algebra 2

Problem Solving

13-6   *The Law of Cosines*

Standing on a small bluff overlooking a local pond, Clay wants to calculate the width of the pond.

1. From point C, Clay walks the distances $\overline{CA}$ and $\overline{CB}$. Then he measures the angle between these line segments.

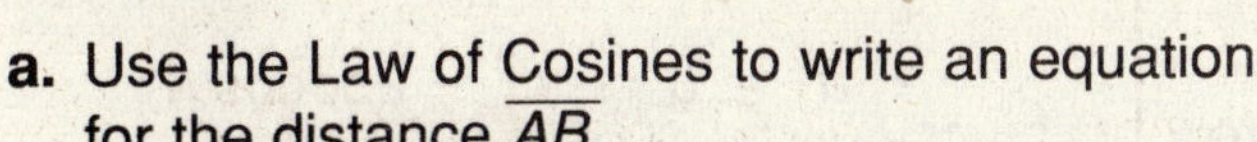

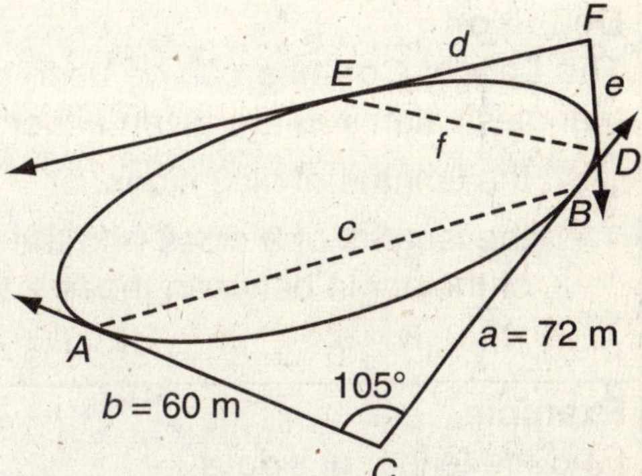

   a. Use the Law of Cosines to write an equation for the distance $\overline{AB}$.

   ____

   b. What is the distance to the nearest meter from A to B?

   ____

2. From another point F, Clay measures 20 meters to D and 50 meters to E. Reece says that last summer this area dried out so much that he could walk the 49 meters from D to E.

   a. Use the Law of Cosines to write an equation for the measure of the angle between $\overline{DF}$ and $\overline{EF}$.

   b. What is the measure of this angle?

3. Reece tells Clay that when the area defined by $\triangle\,DEF$ dries out, it becomes covered with a grass native to this area of the country. Clay wants to know the area of this section.

   a. Use Heron's Formula to write an expression for the area.

   b. Find the area to the nearest tenth of a square meter.

A local naturalist says that a triangular area on the north sides of the pond is a turtle habitat. The lengths of the sides of this area are 15 meters, 25 meters, and 36 meters. Choose the letter for the best answer.

4. What is the area of this triangular turtle habitat?

   A 150.7 m^2

   B 213.2 m^2

   C 1012.9 m^2

   D 3075 m^2

5. Which expression gives the measure of the angle between the sides of the habitat that measure 15 meters and 25 meters?

   F $\cos^{-1}(-0.595)$

   G $\cos(-0.595)$

   H $\cos(0.595)$

   J $\cos^{-1}(0.595)$

Holt Algebra 2

LESSON 13-6 Reading Strategy
Use a Graphic Organizer

A graphic organizer can be used to organize what you know about using
the **Law of Cosines** to solve triangles.

Definition	Use
The **Law of Cosines** can be used to solve triangles when you are given either: • the lengths of all 3 sides, or • the lengths of 2 sides and the measure of the angle between those 2 sides. $$a^2 = b^2 + c^2 - 2bc\cos A$$	This law can be used to solve for any side of a triangle, a, b, or c. $$a^2 = b^2 + c^2 - 2bc\cos A$$ $$b^2 = a^2 + c^2 - 2ac\cos B$$ $$c^2 = a^2 + b^2 - 2ab\cos C$$
Example Find the length of side a. 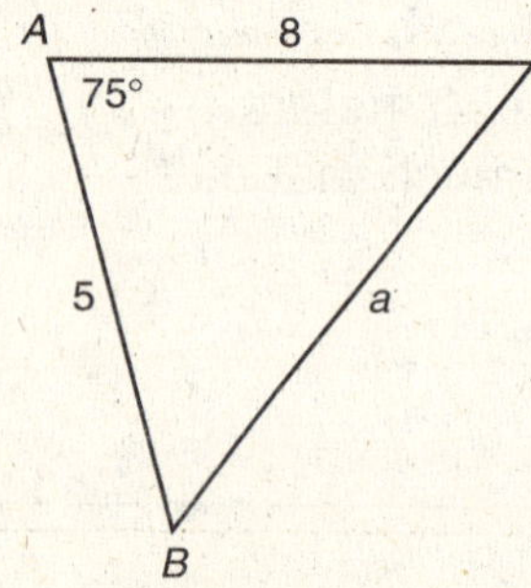 $$a^2 = b^2 + c^2 - 2bc\cos A$$ $$a^2 = 8^2 + 5^2 - 2(8)(5)\cos 75° = 68.3$$ $$a = 8.3$$	**Hint** You can use the Law of Cosines to check results from using the Law of Sines.

Answer each question.

1. Can you find the lengths of all 3 sides of a triangle if you know the measure of 2 angles? Explain.

   ____

2. Describe how you can check your answer when you use the Law of Cosines to solve a triangle.

   ____

3. Jenna is given the following information about triangle *EFG*: $e = 9$ cm, $f = 6$ cm, and $\angle G = 70°$.

   a. Write an equation she can use to find the length of side g.

      ____

   b. Once she has found the length of side g, describe how she can find the measures of $\angle E$ and $\angle F$.

      ____

Holt Algebra 2

Practice A
Right-Angle Trigonometry

Find the value of the sine, cosine, and tangent functions for θ.

1. 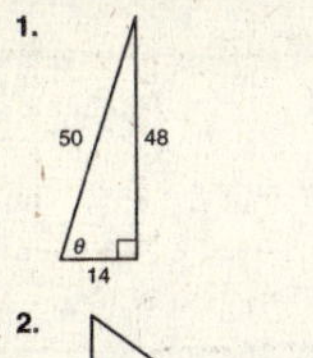

Write and simplify the fractions.

a. $\sin\theta = \dfrac{\text{opp.}}{\text{hyp.}} =$ $\quad \sin\theta = \dfrac{48}{50} = \dfrac{24}{25}$

b. $\cos\theta = \dfrac{\text{adj.}}{\text{hyp.}} =$ $\quad \cos\theta = \dfrac{14}{50} = \dfrac{7}{25}$

c. $\tan\theta = \dfrac{\text{opp.}}{\text{adj.}} =$ $\quad \tan\theta = \dfrac{48}{14} = \dfrac{24}{7}$

2. 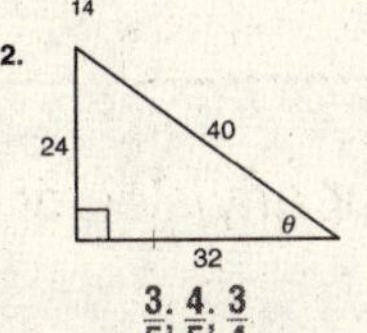

$\dfrac{3}{5}, \dfrac{4}{5}, \dfrac{3}{4}$

3. 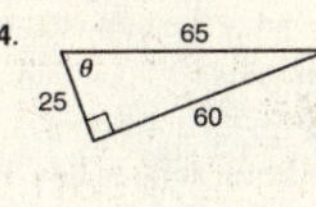

$\dfrac{9}{41}, \dfrac{40}{41}, \dfrac{9}{40}$

4. 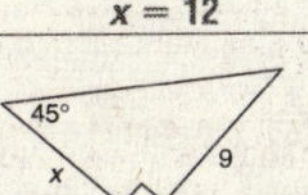

$\dfrac{12}{13}, \dfrac{5}{13}, \dfrac{12}{5}$

Use a trigonometric function to find the value of x.

5. 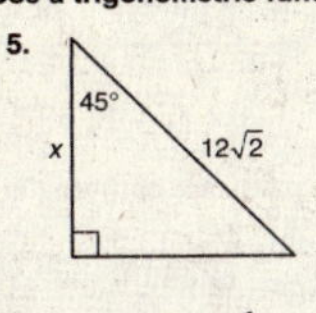

a. Choose the function.

b. Substitute given values.

c. Evaluate the trigonometric ratio for the angle measure.

d. Solve the proportion for x.

Cosine

$\cos 45° = \dfrac{x}{12\sqrt{2}}$

$\cos 45° = \dfrac{\sqrt{2}}{2}$

$x = 12$

6. 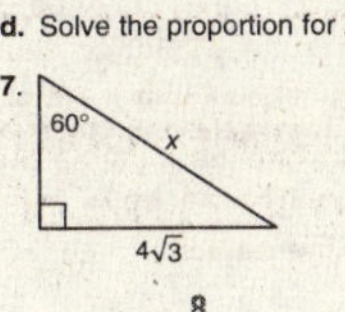

10

7.

8

8.

9

Solve.

9. A conveyor belt leads from the ground to a barn door 24 feet high. The angle between the belt and the ground is 32°. What is the length of the conveyor belt to the nearest foot?

45 ft

Practice B
Right-Angle Trigonometry

Find the value of the sine, cosine, and tangent functions for θ.

1. 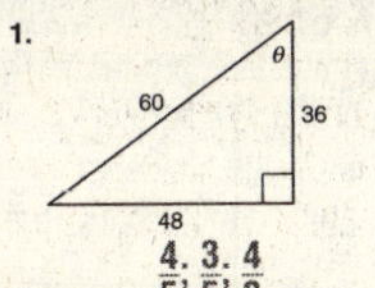

$\dfrac{4}{5}, \dfrac{3}{5}, \dfrac{4}{3}$

2. 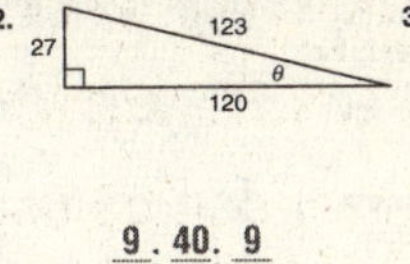

$\dfrac{9}{41}, \dfrac{40}{41}, \dfrac{9}{40}$

3. 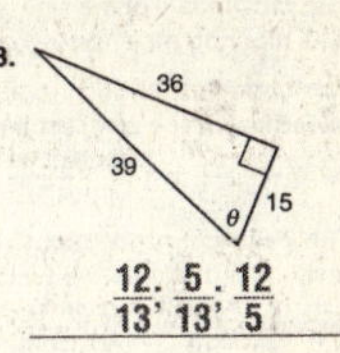

$\dfrac{12}{13}, \dfrac{5}{13}, \dfrac{12}{5}$

Use a trigonometric function to find the value of x.

4. 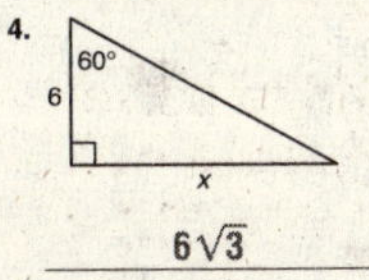

$6\sqrt{3}$

5. 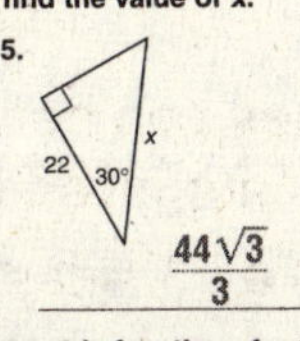

$\dfrac{44\sqrt{3}}{3}$

6. 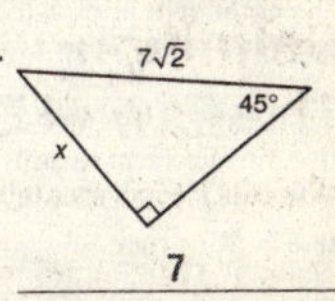

7

Find the values of the six trigonometric functions for θ.

7. 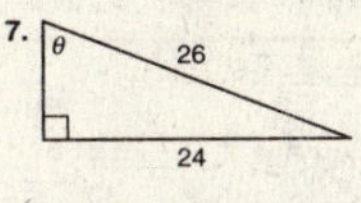

$\sin\theta = \dfrac{12}{13}; \cos\theta = \dfrac{5}{13}; \tan\theta = \dfrac{12}{5}$

$\csc\theta = \dfrac{13}{12}; \sec\theta = \dfrac{13}{5}; \cot\theta = \dfrac{5}{12}$

8. 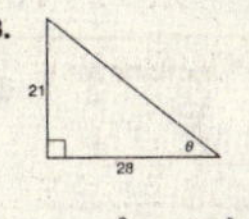

$\sin\theta = \dfrac{3}{5}; \cos\theta = \dfrac{4}{5}; \tan\theta = \dfrac{3}{4}$

$\csc\theta = \dfrac{5}{3}; \sec\theta = \dfrac{5}{4}; \cot\theta = \dfrac{4}{3}$

9. 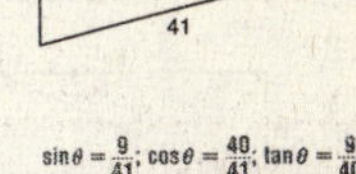

$\sin\theta = \dfrac{9}{41}; \cos\theta = \dfrac{40}{41}; \tan\theta = \dfrac{9}{40}$

$\csc\theta = \dfrac{41}{9}; \sec\theta = \dfrac{41}{40}; \cot\theta = \dfrac{40}{9}$

Solve.

10. A water slide is 26 feet high. The angle between the slide and the water is 33.5°. What is the length of the slide?

47 ft

11. A surveyor stands 150 feet from the base of a viaduct and measures the angle of elevation to be 46.2°. His eye level is 6 feet above the ground. What is the height of the viaduct to the nearest foot?

162 ft

12. The pilot of a helicopter measures the angle of depression to a landing spot to be 18.8°. If the pilot's altitude is 1640 meters, what is the horizontal distance to the landing spot to the nearest meter?

4817 m

Practice C
Right-Angle Trigonometry

Find the value of the sine, cosine, and tangent functions for θ.

1. 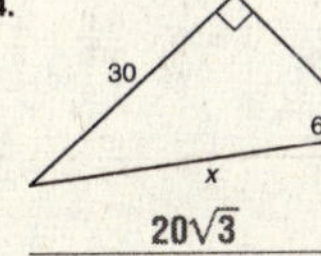

$\dfrac{7}{25}, \dfrac{24}{25}, \dfrac{7}{24}$

2. 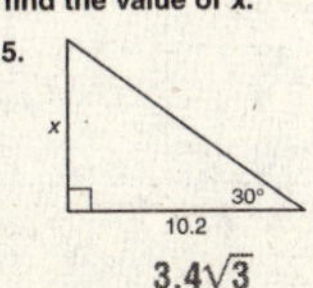

$\dfrac{4}{5}, \dfrac{3}{5}, \dfrac{4}{3}$

3. 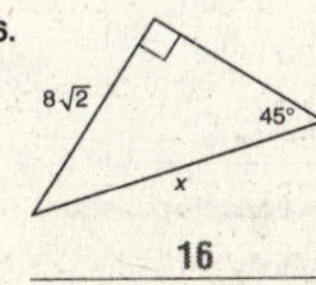

$\dfrac{60}{61}, \dfrac{11}{61}, \dfrac{60}{11}$

Use a trigonometric function to find the value of x.

4.

$20\sqrt{3}$

5.

$3.4\sqrt{3}$

6. 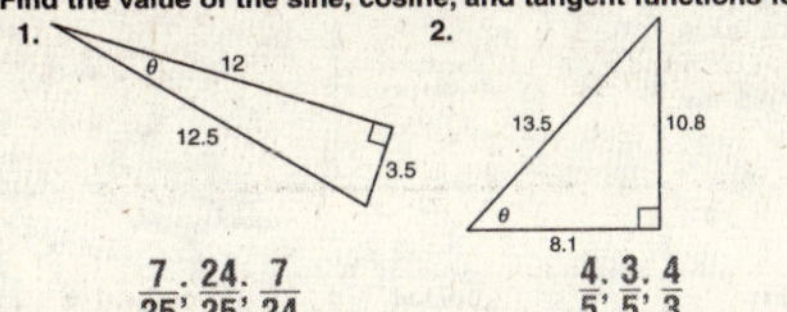

16

Find the values of the six trigonometric functions for θ.

7. 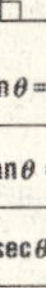

$\sin\theta = \dfrac{7}{25}; \cos\theta = \dfrac{24}{25};$

$\tan\theta = \dfrac{7}{24}; \csc\theta = \dfrac{25}{7};$

$\sec\theta = \dfrac{25}{24}; \cot\theta = \dfrac{24}{7}$

8.

$\sin\theta = \dfrac{12}{13}; \cos\theta = \dfrac{5}{13};$

$\tan\theta = \dfrac{12}{5}; \csc\theta = \dfrac{13}{12};$

$\sec\theta = \dfrac{13}{5}; \cot\theta = \dfrac{5}{12}$

9.

$\sin\theta = \dfrac{\sqrt{10}}{5}; \cos\theta = \dfrac{\sqrt{15}}{5};$

$\tan\theta = \dfrac{\sqrt{6}}{3}; \csc\theta = \dfrac{\sqrt{10}}{2};$

$\sec\theta = \dfrac{\sqrt{15}}{3}; \cot\theta = \dfrac{\sqrt{6}}{2}$

Solve.

10. A kite string is 102 feet long. The angle between the kite string and the ground is 54.9°. How high is the kite?

83.5 ft

11. A surveyor stands 186 feet from the base of a cliff and measures the angle of elevation to be 56.6°. His eye level is 5 feet above the ground. What is the height of the cliff to the nearest foot?

287 ft

12. The pilot of a hot air balloon measures the angle of depression to a landing spot to be 36.7°. If the pilot's altitude is 1752 meters, what is the horizontal distance to the landing spot to the nearest meter?

2350 m

Reteach
Right-Angle Trigonometry

A **trigonometric ratio** compares the lengths of two sides of a right triangle. The values of the ratios depend upon one of the acute angles of the triangle, denoted by θ.

> **S**ine is **O**pposite over **H**ypotenuse,
> **C**osine is **A**djacent over **H**ypotenuse,
> **T**angent is **O**pposite over **A**djacent.

Use **SOHCAHTOA** to remember the relationships between the sides of a right triangle that correspond to the trigonometric ratios sine, cosine, and tangent.

$\sin\theta = \dfrac{\text{Opposite}}{\text{Hypotenuse}} = \dfrac{a}{c}$

$\cos\theta = \dfrac{\text{Adjacent}}{\text{Hypotenuse}} = \dfrac{b}{c}$

$\tan\theta = \dfrac{\text{Opposite}}{\text{Adjacent}} = \dfrac{a}{b}$

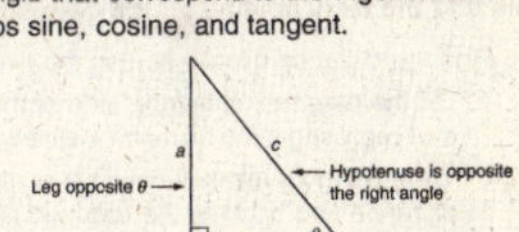

Use the definitions of each ratio and the corresponding values from a given right triangle to find the values of the trigonometric functions for θ.

$\sin\theta = \dfrac{\text{Opposite}}{\text{Hypotenuse}} = \dfrac{14}{50} = \dfrac{7}{25}$

$\cos\theta = \dfrac{\text{Adjacent}}{\text{Hypotenuse}} = \dfrac{48}{50} = \dfrac{24}{25}$

$\tan\theta = \dfrac{\text{Opposite}}{\text{Adjacent}} = \dfrac{14}{48} = \dfrac{7}{24}$

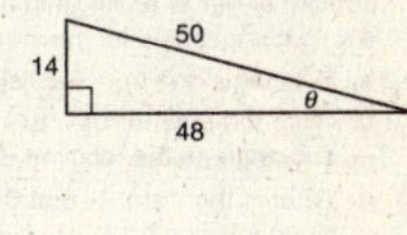

Find the value of the sine, cosine, and tangent functions for θ.

1. 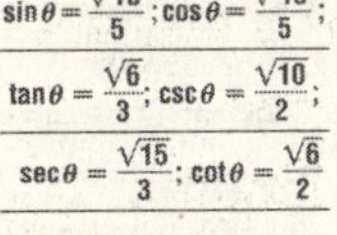

2. 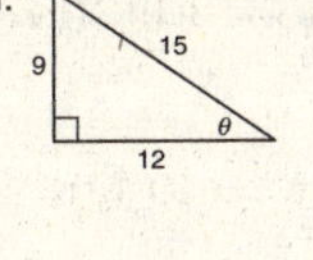

3. 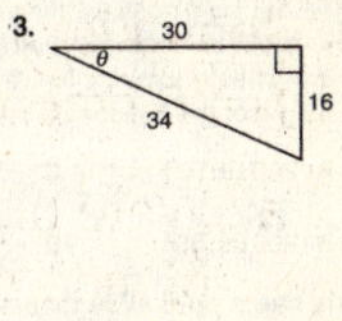

$\sin\theta = \dfrac{\text{Opposite}}{\text{Hypotenuse}} = \dfrac{3}{5}$

$\cos\theta = \dfrac{\text{Adjacent}}{\text{Hypotenuse}} = \dfrac{4}{5}$

$\tan\theta = \dfrac{\text{Opposite}}{\text{Adjacent}} = \dfrac{3}{4}$

$\sin\theta = \dfrac{12}{13}$

$\cos\theta = \dfrac{5}{13}$

$\tan\theta = \dfrac{12}{5}$

$\sin\theta = \dfrac{8}{17}$

$\cos\theta = \dfrac{15}{17}$

$\tan\theta = \dfrac{8}{15}$

 Reteach
Right-Angle Trigonometry (continued)

The reciprocals of the sine, cosine, and tangent ratios are also trigonometric ratios.

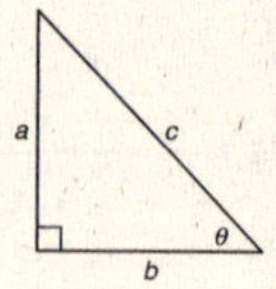

The **cosecant** function (**cscθ**) is the reciprocal of the sine function.
$$\csc\theta = \frac{1}{\sin\theta} = \frac{\text{Hypotenuse}}{\text{Opposite}} = \frac{c}{a}$$

The **secant** function (**secθ**) is the reciprocal of the cosine function.
$$\sec\theta = \frac{1}{\cos\theta} = \frac{\text{Hypotenuse}}{\text{Adjacent}} = \frac{c}{b}$$

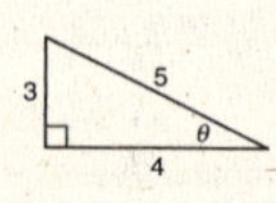

The **cotangent** function (**cotθ**) is the reciprocal of the tangent function.
$$\cot\theta = \frac{1}{\tan\theta} = \frac{\text{Adjacent}}{\text{Opposite}} = \frac{b}{a}$$

Use the reciprocal relationship of the ratios to find the values of the reciprocal trigonometric functions.

$$\sin\theta = \frac{3}{5} \qquad \csc\theta = \frac{1}{\sin\theta} = \frac{5}{3}$$
$$\cos\theta = \frac{4}{5} \qquad \sec\theta = \frac{1}{\cos\theta} = \frac{5}{4}$$
$$\tan\theta = \frac{3}{4} \qquad \cot\theta = \frac{1}{\tan\theta} = \frac{4}{3}$$

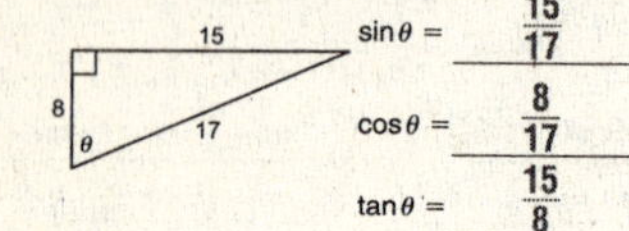

Find the values of the six trigonometric functions for θ.

4.

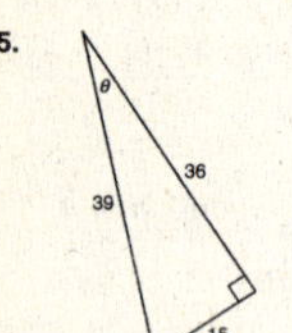

$$\sin\theta = \frac{15}{17} \qquad \csc\theta = \frac{1}{\sin\theta} = \frac{17}{15}$$
$$\cos\theta = \frac{8}{17} \qquad \sec\theta = \frac{1}{\cos\theta} = \frac{17}{8}$$
$$\tan\theta = \frac{15}{8} \qquad \cot\theta = \frac{1}{\tan\theta} = \frac{8}{15}$$

5.

$$\sin\theta = \frac{5}{13} \qquad \csc\theta = \frac{13}{5}$$
$$\cos\theta = \frac{12}{13} \qquad \sec\theta = \frac{13}{12}$$
$$\tan\theta = \frac{5}{12} \qquad \cot\theta = \frac{12}{5}$$

7   **Holt Algebra 2**

 Challenge
Twice Right

Trigonometry can be used to solve problems involving one or more right triangles. Different triangles can share common parts.

As shown in the diagram at right, a pole, $\overline{TF}$, is on the roof of a shed, $\overline{FB}$. From a point, P, on the ground 27 feet from the foot of the shed, the measure of the angle of elevation to the top of the pole, T, is 38°, and the measure of the angle of elevation to the foot of the pole, F, is 32°. Determine, to the nearest tenth of a foot, the height of the pole.

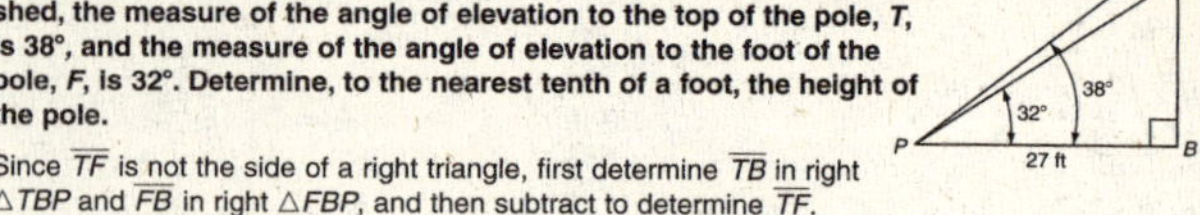

Since $\overline{TF}$ is not the side of a right triangle, first determine $\overline{TB}$ in right $\triangle TBP$ and $\overline{FB}$ in right $\triangle FBP$, and then subtract to determine $\overline{TF}$.

In right $\triangle TBP$:
$$\tan 38° = \frac{\text{opp.}}{\text{adj.}} = \frac{TB}{27}$$
$$\overline{TB} = 27\tan 38° \approx 21.09$$

In right $\triangle FBP$:
$$\tan 32° = \frac{\text{opp.}}{\text{adj.}} = \frac{FB}{27}$$
$$\overline{FB} = 27\tan 32° \approx 16.87$$

So, $\overline{TF} = \overline{TB} - \overline{FB} \approx 21.09 - 16.87 \approx 4.2$ feet.

As shown in the diagram at right, a ship is headed directly toward a coastline formed by a vertical cliff ($\overline{BC}$) that is 80 meters high. At point A, from the ship, the measure of the angle of elevation to the top of the cliff (B) is 10°. A few minutes later, at point D, the measure of the angle of elevation to the top of the cliff has increased to 20°. Determine the following to the nearest meter.

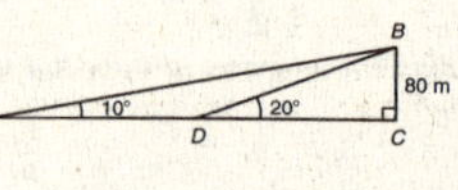

1. $\overline{DC}$ **220 m**

2. $\overline{AC}$ **454 m**

3. The distance between the 2 sightings   **234 m**

As shown in the diagram at right, the heights ($\overline{AB}$ and $\overline{CD}$) of buildings on opposite sides of a 90-foot-wide avenue are, respectively, 100 feet and 50 feet. From a point E on the avenue, the measure of the angle of elevation to B is 55°. Determine each of the following.

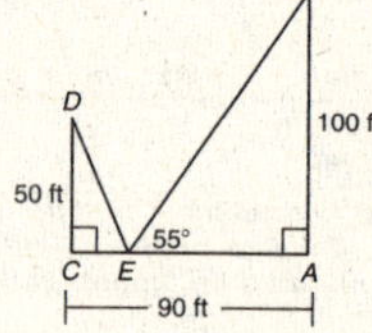

4. The distance from E to A to the nearest foot   **70 ft**

5. The distance from E to D to the nearest foot   **54 ft**

8   **Holt Algebra 2**

 Problem Solving
Right-Angle Trigonometry

Kayla is fishing near the ferry landing near her home. She wonders how far the ferries actually travel when they cross the river. To find out, she puts a fishing pole upright on the riverbank directly across from the ferry landing. Then she walks down the bank 180 yards and measures an angle of 75° between the lines to her fishing pole and the ferry landing on the opposite bank.

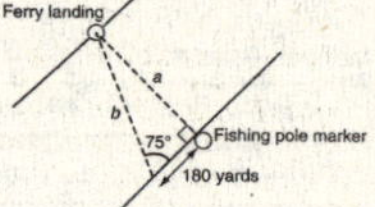

1. Find the distance directly across the river.

   a. On the diagram, name the side of the triangle that represents the distance Kayla wants to determine.   *a*

   b. Write a trigonometric function that relates the known distance and angle to the required distance. $\tan 75° = \dfrac{a}{180}$

   c. Find the distance to the nearest yard directly across the river.   **672 yd**

2. Kayla walks another 200 yards down the riverbank. At this point the ferries seem to come straight at her at an angle of 60° to the line to her fishing pole. If the boats travel along this line to compensate for the current, how far do they travel?

   a. How far is she from her fishing pole now?   **380 yd**

   b. Write the trigonometric function that relates this distance and angle to the required distance, *c*. $\cos 60° = \dfrac{380}{c}$

   c. What is the distance that the ferries travel if they travel along this line?   **760 yd**

At a hot-air balloon festival, Luis watches a hot-air balloon rise from a distance of 200 yards. Choose the letter for the best answer.

3. From Luis's position, the balloon seems to hover at an angle of elevation of 50°. Which trigonometric function gives the height of the balloon, *h*?

   A $200\sin 50°$

   (B) $200\tan 50°$

   C $\dfrac{200}{\cos 50°}$

   D $\dfrac{200}{\cot 50°}$

4. After a short while, the balloon seems to hover at an angle of 75°. How high is it off the ground now?

   F 193 yd

   G 207 yd

   (H) 746 yd

   J 773 yd

Olivia has a pool slide that makes an angle of 25° with the water. The top of the slide stands 4.5 feet above the surface of the water. Choose the letter for the best answer.

5. How far out into the pool will the slide reach?

   A 2.1 ft

   B 5.0 ft

   C 7.6 ft

   (D) 9.7 ft

6. The slide makes a straight line into the water. How long is the slide?

   F 5.0 ft

   G 7.6 ft

   H 9.7 ft

   (J) 10.6 ft

9   **Holt Algebra 2**

 Reading Strategy
Identify Relationships

Trigonometric functions are used to describe the relationship between side length and angle measurements of right triangles. Part of finding the functions involves understanding these relationships.

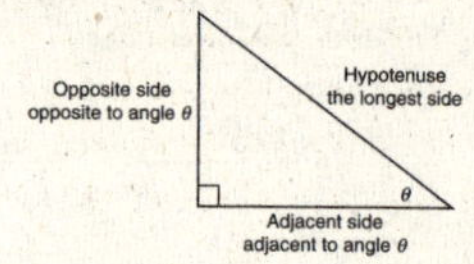

Triangle	sine	cosine	tangent
	$\sin\theta = \dfrac{\text{opposite}}{\text{hypotenuse}}$	$\cos\theta = \dfrac{\text{adjacent}}{\text{hypotenuse}}$	$\tan\theta = \dfrac{\text{opposite}}{\text{adjacent}}$
	$\sin\theta = \dfrac{\text{opp.}}{\text{hyp.}} = \dfrac{4}{5}$	$\cos\theta = \dfrac{\text{adj.}}{\text{hyp.}} = \dfrac{3}{5}$	$\tan\theta = \dfrac{\text{opp.}}{\text{adj.}} = \dfrac{4}{3}$

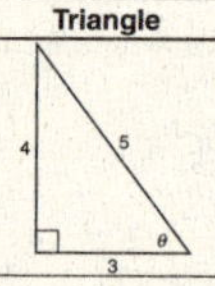

Answer each question.

1. For triangle *FGH*

   a. What is the length of the side opposite θ?   **5**

   b. What is the length of the side adjacent to θ?   **8**

   c. What is the length of the hypotenuse?   **9.95**

   d. Write a fraction to represent cosine θ. $\dfrac{8}{9.95}$

   e. Write a fraction to represent tangent θ. $\dfrac{5}{8}$

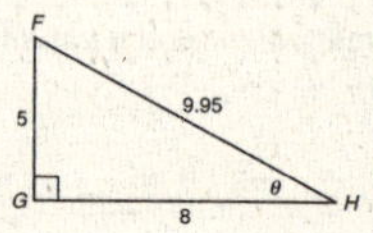

2. For triangle *JKL*

   a. What is the length of the side opposite θ?   **12**

   b. What is the length of the side adjacent to θ?   **9**

   c. What is the length of the hypotenuse?   **15**

   d. Write a fraction to represent cosine θ. $\dfrac{9}{15}$ or $\dfrac{3}{5}$

   e. Write a fraction to represent tangent θ. $\dfrac{12}{9}$ or $\dfrac{4}{3}$

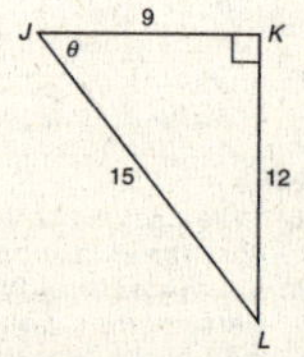

3. Compare the sine of the two acute angles in an isosceles right triangle. Explain.

   An isosceles triangle has 2 sides that are equal in length. Since the hypotenuse is the longest side of a right triangle, the equal sides must be the 2 legs. Therefore, the 2 angles are the same. Since the angles are equal, both 45°, they have the same sine.

10   **Holt Algebra 2**

    Holt Algebra 2

Practice A
Angles of Rotation

Draw an angle with the given measure in standard position.

1. −390°

The angle is negative, so rotate clockwise from 0°.

2. 315°

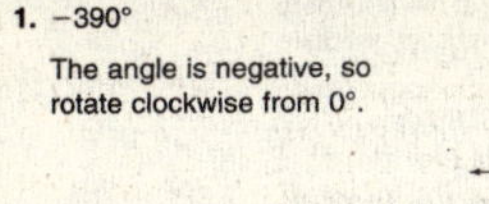
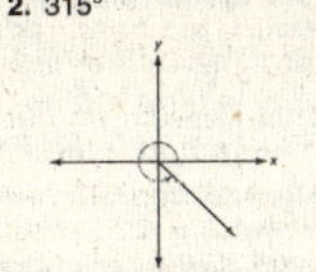

3. −120°   **4.** 240°   **5.** −585°

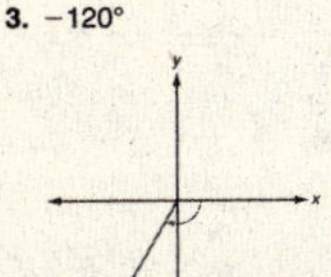
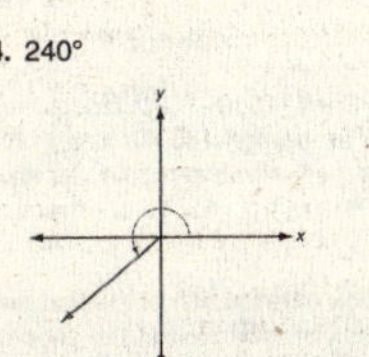
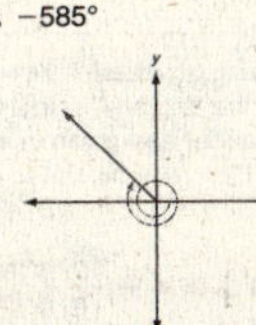

Find the measures of a positive angle and a negative angle that are coterminal with each given angle.

6. $\theta = 28°$
a. Add 360°. **388°**
b. Subtract 360°. **−332°**

7. $\theta = 250°$
a. Add 360°. **610°**
b. Subtract 360°. **−110°**

8. $\theta = 740°$ **20°, −340°**

9. $\theta = -6°$ **354°, −366°**

10. $\theta = -370°$ **350°, −10°**

11. $\theta = 515°$ **155°, −205°**

12. $\theta = -93°$ **267°, −453°**

13. $\theta = 162°$ **522°, −198°**

Find the measure of the reference angle for each given angle.

14. $\theta = 475°$ **65°**

15. $\theta = 212°$ **32°**

16. $\theta = -115°$ **65°**

17. $\theta = 740°$ **20°**

18. $\theta = -96°$ **84°**

19. $\theta = -401°$ **41°**

20. $\theta = 320°$ **40°**

21. $\theta = -722°$ **2°**

22. $\theta = 292°$ **68°**

23. $\theta = -850°$ **50°**

24. $\theta = 1000°$ **80°**

25. $\theta = -1000°$ **80°**

11
Holt Algebra 2

Practice B
Angles of Rotation

Draw an angle with the given measure in standard position.

1. −420°   **2.** 405°   **3.** −450°

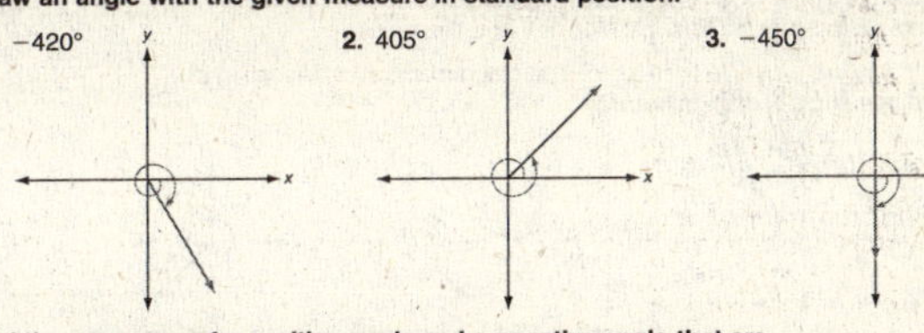

Find the measures of a positive angle and a negative angle that are coterminal with each given angle.

4. $\theta = 425°$ **65°, −295°**

5. $\theta = -316°$ **44°, −676°**

6. $\theta = -800°$ **280°, −80°**

7. $\theta = 281°$ **641°, −79°**

8. $\theta = -4°$ **356°, −364°**

9. $\theta = 743°$ **23°, −337°**

Find the measure of the reference angle for each given angle.

10. $\theta = 211°$ **31°**

11. $\theta = -755°$ **35°**

12. $\theta = -555°$ **15°**

13. $\theta = 119°$ **61°**

14. $\theta = -160°$ **20°**

15. $\theta = 235°$ **55°**

P is a point on the terminal side of θ in standard position. Find the exact value of the six trigonometric functions for θ.

16. $P(-5, 5)$

$\sin\theta = \frac{\sqrt{2}}{2}; \cos\theta = -\frac{\sqrt{2}}{2};$

$\tan\theta = -1; \cot\theta = -1;$

$\csc\theta = \sqrt{2}; \sec\theta = -\sqrt{2}$

17. $P(2, 9)$

$\sin\theta = \frac{9\sqrt{85}}{85}; \cos\theta = \frac{2\sqrt{85}}{85};$

$\tan\theta = \frac{9}{2}; \cot\theta = \frac{2}{9};$

$\csc\theta = \frac{\sqrt{85}}{9}; \sec\theta = \frac{\sqrt{85}}{2}$

18. $P(-7, -5)$

$\sin\theta = -\frac{5\sqrt{74}}{74}; \cos\theta = -\frac{7\sqrt{74}}{74}$

$\tan\theta = \frac{5}{7}; \cot\theta = \frac{7}{5};$

$\csc\theta = -\frac{\sqrt{74}}{5}; \sec\theta = -\frac{\sqrt{74}}{7}$

Solve.

19. A circus performer trots her pony into the ring. The pony circles the ring 22 times as the performer flips and turns on the pony's back. At the end of the act, the pony exits on the side of the ring opposite its point of entry. Through how many degrees does the pony trot during the entire act?   **8100°**

12
Holt Algebra 2

Practice C
Angles of Rotation

Find the measures of a positive angle and a negative angle that are coterminal with each given angle.

1. $\theta = 400°$ **40°, −320°**

2. $\theta = -360°$ **360°, −720°**

3. $\theta = -1010°$ **70°, −290°**

4. $\theta = 567°$ **207°, −153°**

5. $\theta = -164°$ **196°, −524°**

6. $\theta = 358°$ **718°, −2°**

Find the measure of the reference angle for each given angle.

7. $\theta = 504°$ **36°**

8. $\theta = -388°$ **28°**

9. $\theta = 991°$ **91°**

10. $\theta = 486°$ **54°**

11. $\theta = -920°$ **20°**

12. $\theta = -1787°$ **13°**

P is a point on the terminal side of θ in standard position. Find the exact value of the six trigonometric functions for θ.

13. $P(-1, 14)$

$\sin\theta = \frac{14\sqrt{197}}{197}; \cos\theta = -\frac{\sqrt{197}}{197};$

$\tan\theta = -14; \cot\theta = -\frac{1}{14};$

$\csc\theta = \frac{\sqrt{197}}{14}; \sec\theta = -\sqrt{197}$

14. $P(-8, -8)$

$\sin\theta = -\frac{\sqrt{2}}{2}; \cos\theta = -\frac{\sqrt{2}}{2};$

$\tan\theta = 1; \cot\theta = 1;$

$\csc\theta = -\sqrt{2}; \sec\theta = -\sqrt{2}$

15. $P(9, -6)$

$\sin\theta = -\frac{2\sqrt{13}}{13}; \cos\theta = \frac{3\sqrt{13}}{13};$

$\tan\theta = -\frac{2}{3}; \cot\theta = -\frac{3}{2};$

$\csc\theta = -\frac{\sqrt{13}}{2}; \sec\theta = \frac{\sqrt{13}}{3}$

16. $P(10, 15)$

$\sin\theta = \frac{3\sqrt{13}}{13}; \cos\theta = \frac{2\sqrt{13}}{13};$

$\tan\theta = \frac{3}{2}; \cot\theta = \frac{2}{3};$

$\csc\theta = \frac{\sqrt{13}}{3}; \sec\theta = \frac{\sqrt{13}}{2}$

17. $P(-2, -1)$

$\sin\theta = -\frac{\sqrt{5}}{5}; \cos\theta = -\frac{2\sqrt{5}}{5};$

$\tan\theta = \frac{1}{2}; \cot\theta = 2;$

$\csc\theta = -\sqrt{5}; \sec\theta = -\frac{\sqrt{5}}{2}$

18. $P(-12, 5)$

$\sin\theta = \frac{5}{13}; \cos\theta = -\frac{12}{13};$

$\tan\theta = -\frac{5}{12}; \cot\theta = -\frac{12}{5};$

$\csc\theta = \frac{13}{5}; \sec\theta = -\frac{13}{12}$

Solve.

19. A restaurant in the round rotates clockwise so diners can view the city. Fifty evenly-spaced window tables are numbered clockwise from 1 to 50. A waiter noted where Table 1 was at the beginning of his shift. At the end of his shift, the restaurant had made 4 complete rotations and Table 1 was then where Table 22 had been. Through how many degrees had the restaurant rotated during his shift?   **1598.4°**

13
Holt Algebra 2

Reteach
Angles of Rotation

To draw an angle with a given measure, start with the initial side on the x-axis and rotate the terminal side to show the angle measure.

- Rotate *counterclockwise* for a positive angle measure.
- Rotate *clockwise* for a negative angle measure.

> Label the angle measures on the axes to help locate the correct quadrant.

Draw a 210° angle.

Start at the x-axis and rotate counterclockwise past 180° to 210°. Draw the terminal side. Label the angle.

Draw a −145° angle.

Start at the x-axis and rotate clockwise past −90° to −145°. Draw the terminal side. Label the angle.

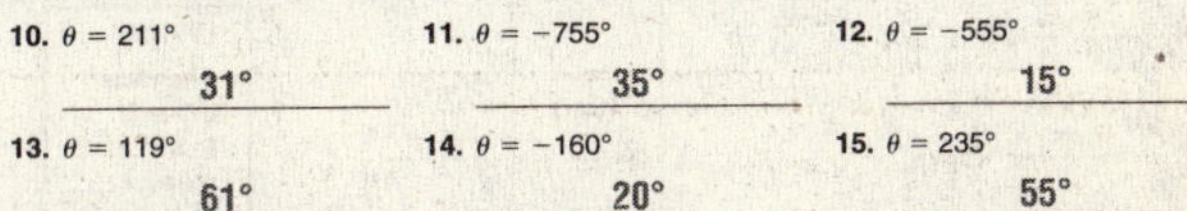

Coterminal angles are angles with the same initial and terminal sides. You can find the measure of an angle that is coterminal with another angle by adding 360° to or subtracting 360° from the angle.

For $\theta = 65°$:

$65° + 360° = 425°$

$65° - 360° = -295°$

> 360° is a complete rotation.

425° and −295° are coterminal with 65°.

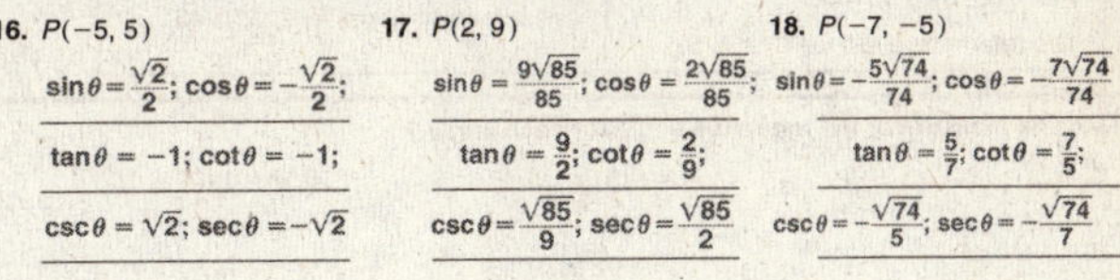

Draw an angle with the given measure in standard position.

1. 150°   **2.** −330°

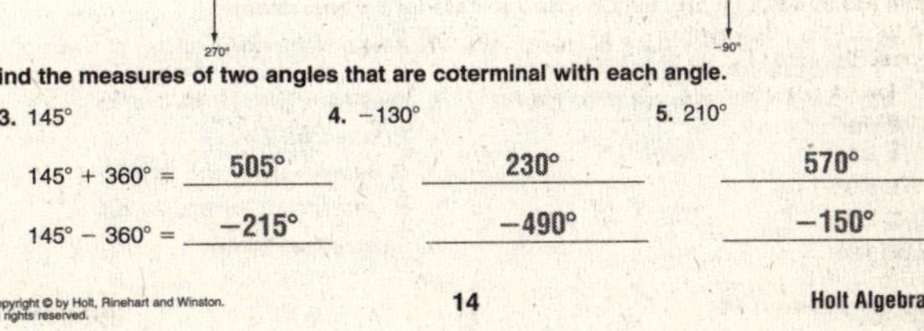

Find the measures of two angles that are coterminal with each angle.

3. 145°

$145° + 360° =$ **505°**

$145° - 360° =$ **−215°**

4. −130°   **230°**   **−490°**

5. 210°   **570°**   **−150°**

14
Holt Algebra 2

Holt Algebra 2

Reteach
Angles of Rotation (continued)

The **reference angle** for an angle θ in standard position is the positive acute angle formed by the terminal side of θ and the x-axis.

Denote the reference angle for θ as θ'. The diagram shows the location of the reference angle in each quadrant.

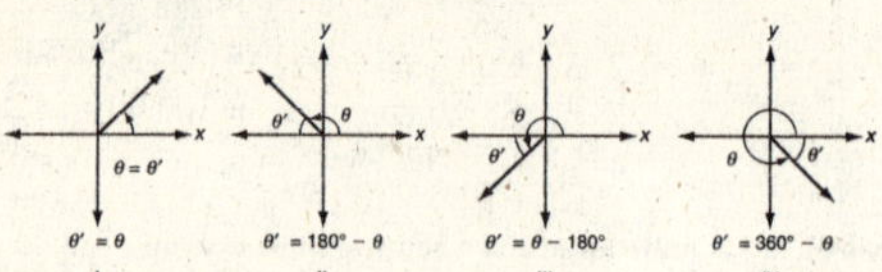

To find the reference angle for an angle θ:

- draw the angle in standard position,
- find the reference angle formed by the terminal side of θ and the x-axis,
- find the measure of θ'.

θ' is an acute angle so it is always less than 90°.

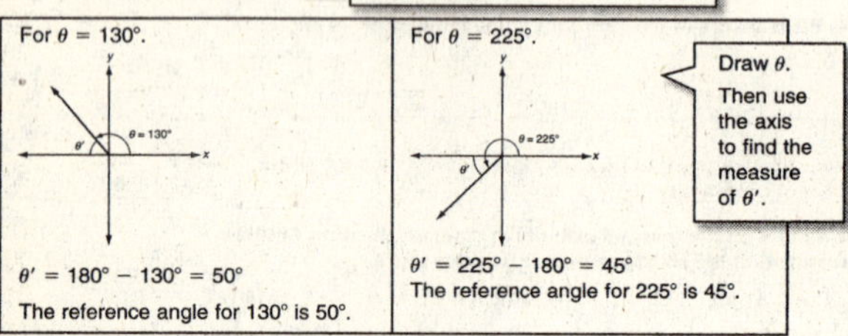

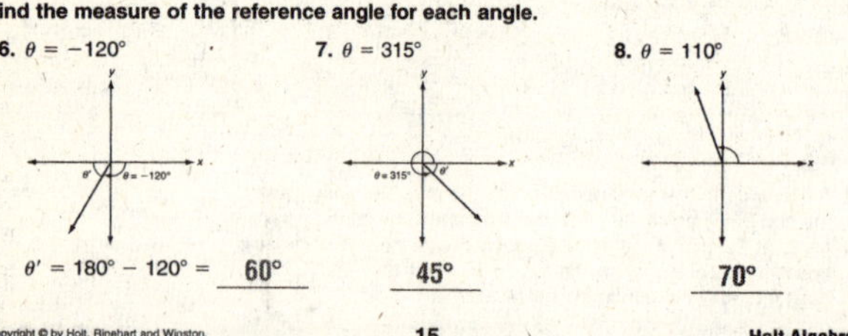

Draw θ. Then use the axis to find the measure of θ'.

$\theta' = 180° - 130° = 50°$
The reference angle for 130° is 50°.

$\theta' = 225° - 180° = 45°$
The reference angle for 225° is 45°.

Find the measure of the reference angle for each angle.

6. $\theta = -120°$

7. $\theta = 315°$

8. $\theta = 110°$

$\theta' = 180° - 120° = \quad$ **60°**

45°

70°

15   **Holt Algebra 2**

Challenge
Exploring the Behavior of Light

When a light ray passes from one medium, such as air, into another medium, such as glass, the light ray is bent, or *refracted*, at the boundary between the two media. The *index of refraction*, n, of a particular medium is given by the equation below.

$$n = \frac{\sin\theta_i}{\sin\theta_r} \quad \begin{cases} \theta_i \text{ is the incoming } \textit{angle of incidence} \text{ of the light ray} \\ \theta_r \text{ is the outgoing } \textit{angle of refraction} \end{cases}$$

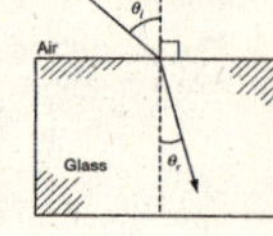

1. Measurements taken using a piece of crown glass gave an angle of incidence of 43.7° and an angle of refraction of 27.1°. Determine the index of refraction of crown glass.

About 1.52

The Dutch physicist Willebrond Snell (1580–1626) discovered that the sine of the angle of incidence is proportional to the sine of the angle of refraction for any given interface between two media. As shown below, Snell's Law may be stated in terms of the indices of refraction of the media.

$$n_1 \sin\theta_i = n_2 \sin\theta_r \quad \begin{cases} n_1 \text{ is the index of refraction of the first medium} \\ n_2 \text{ is the index of refraction of the second medium} \end{cases}$$

2. The measure of θ_i is 30° and the measure of θ_r for crown glass is 19.2°. Use the index of refraction of crown glass that you determined in Exercise 1. Find the index of refraction of air.

1.0

Snell's Law may be also stated in terms of the velocity of light in the two media.

$$\frac{\sin\theta_i}{\sin\theta_r} = \frac{v_1}{v_2} \quad \begin{cases} v_1 \text{ is the speed of light in the first medium} \\ v_2 \text{ is the speed of light in the second medium} \end{cases}$$

3. The measure of the angle of incidence with which a ray of light strikes crown glass is 30° and the measure of the angle of refraction is 19.2°. The speed of light in air is 3.0×10^8 meters per second. Find the speed of light in crown glass.

1.97×10^8 m per s

When light crosses a boundary and its speed increases in the second medium, the measure of the angle of refraction is greater than the measure of the angle of incidence. There is an angle of incidence, called the *critical angle* (θ_c), for which the corresponding angle of refraction is 90°, and the ray of light emerges parallel to the boundary. You can determine the measure of a critical angle when light passes from some medium into air by using the formula below.

$\sin\theta_c = \frac{1}{n}$, where n is the index of refraction of the first medium

4. When light passes from diamond to air the measure of the critical angle is 24.4°. Find the index of refraction of diamond.

2.42

. 16   **Holt Algebra 2**

Problem Solving
Angles of Rotation

Isabelle and Karl agreed to meet at the rotating restaurant at the top of a tower in the town center. The restaurant makes one full rotation each hour.

1. Isabelle waits for Karl, who arrives 30 minutes later. Through how many degrees does the restaurant rotate between the time that Isabelle arrives and the time that Karl arrives? **180°**

2. Since the restaurant is busy, they don't get menus for another ten minutes. How much farther has the restaurant rotated? **60°**

3. By the time they are served dinner, the restaurant has rotated to an angle 30° short of its orientation when Isabelle arrived.

   a. Write an expression for the length of time that Isabelle has been there. $\dfrac{330}{360} \times 60$ min

   b. How long has she been there? **55 min**

   c. How long has Karl been there? **25 min**

   d. How far has the restaurant rotated since Karl arrived? **150°**

4. When their bill comes, it includes a note that the restaurant has rotated 840° since Isabelle arrived.

   a. How long has it been since they were served dinner? **2 h 20 min**

   b. How many rotations has the restaurant made since Isabelle arrived? $2\frac{1}{3}$

   c. How far is the restaurant from its orientation when Karl arrived? **60°**

5. On their way out, they stop and look at a map. A museum is located at a point northeast of the tower with coordinates (3, 2).

   a. Write the trigonometric function for the angle that the line from the tower to the museum makes with a line due north from the tower. $\tan\theta = \dfrac{3}{2}$

   b. Isabelle says that the museum is exactly 4 kilometers from the tower. How much farther north is the museum than the tower? **2.2 km**

An advertisement on a kiosk near the bus stop rotates through 765° while Aaron waits for his bus. Choose the letter for the best answer.

6. What is the difference in the orientation of the advertisement between when Aaron arrived at the bus stop and when the bus came?
   A 45°
   B 90°
   C 315°
   D 405°

7. If the advertisement rotates at a rate of one rotation every 10 minutes, how long does Aaron wait for his bus?
   F Less than 2 min
   G Between 2 min and 10 min
   H Between 10 min and 20 min
   J More than 20 min

17   **Holt Algebra 2**

Reading Strategy
Understand Vocabulary

There are different terms that are used to describe an angle of rotation on the coordinate plane.

Coterminal angles are angles in standard position with the same terminal side. The angle measuring −235° is coterminal with the angle measuring 125°.

This angle is in **standard position**: Its vertex is at the origin and one ray is on the positive x-axis. This is its **terminal side**. It measures 125°.

The **reference angle** is the positive acute angle formed by the terminal side of an angle and the x-axis. This reference angle is 55°.

Answer each question.

1. An angle has a counterclockwise angle of rotation and the terminal side lies in the third quadrant. What do you know about this angle? **Possible answer: The angle is positive and measures between 180° and 270°.**

2. An angle has a clockwise angle of rotation and the terminal side lies in the first quadrant. What do you know about this angle? **Possible answer: The angle is negative and measures between −270° and −360°.**

3. a. Draw a 245° angle in standard position.

   b. What is the measure of a coterminal angle? **−115°**

   c. What is the measure of the reference angle? **65°**

4. a. Draw a −200° angle in standard position.

   b. What is the measure of a coterminal angle? **160°**

   c. What is the measure of the reference angle? **20°**

5. Do all angles have both positive and negative coterminal angles? Explain. **Yes; because you can find coterminal angles by either adding 360° to or subtracting 360° from the measure of the angle**

6. Are reference angles always positive? Explain. **Yes; because by definition, reference angles are the measure of the positive acute angle made by the terminal side of an angle and the x-axis.**

18   **Holt Algebra 2**

Holt Algebra 2

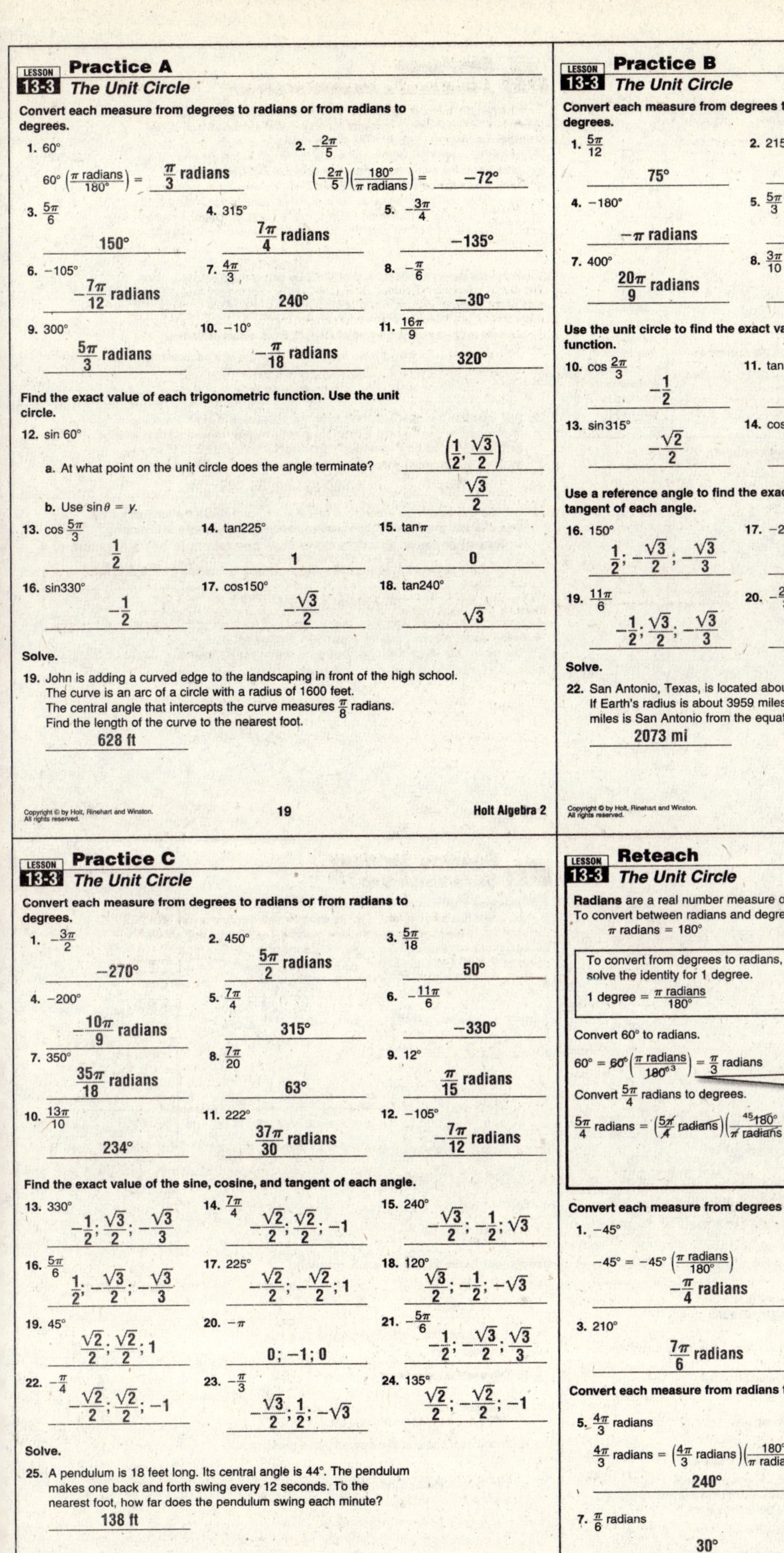

Practice A
13-3 The Unit Circle

Convert each measure from degrees to radians or from radians to degrees.

1. $60°$

$60° \left(\frac{\pi \text{ radians}}{180°}\right) =$ _____ $\frac{\pi}{3}$ radians

2. $-\frac{2\pi}{5}$

$\left(-\frac{2\pi}{5}\right)\left(\frac{180°}{\pi \text{ radians}}\right) =$ _____ $-72°$

3. $\frac{5\pi}{6}$ _____ $150°$

4. $315°$ _____ $\frac{7\pi}{4}$ radians

5. $-\frac{3\pi}{4}$ _____ $-135°$

6. $-105°$ _____ $-\frac{7\pi}{12}$ radians

7. $\frac{4\pi}{3}$ _____ $240°$

8. $-\frac{\pi}{6}$ _____ $-30°$

9. $300°$ _____ $\frac{5\pi}{3}$ radians

10. $-10°$ _____ $-\frac{\pi}{18}$ radians

11. $\frac{16\pi}{9}$ _____ $320°$

Find the exact value of each trigonometric function. Use the unit circle.

12. $\sin 60°$

  a. At what point on the unit circle does the angle terminate? _____ $\left(\frac{1}{2}, \frac{\sqrt{3}}{2}\right)$

  b. Use $\sin\theta = y$. _____ $\frac{\sqrt{3}}{2}$

13. $\cos\frac{5\pi}{3}$ _____ $\frac{1}{2}$

14. $\tan 225°$ _____ 1

15. $\tan\pi$ _____ 0

16. $\sin 330°$ _____ $-\frac{1}{2}$

17. $\cos 150°$ _____ $-\frac{\sqrt{3}}{2}$

18. $\tan 240°$ _____ $\sqrt{3}$

Solve.

19. John is adding a curved edge to the landscaping in front of the high school. The curve is an arc of a circle with a radius of 1600 feet. The central angle that intercepts the curve measures $\frac{\pi}{8}$ radians. Find the length of the curve to the nearest foot.

_____ 628 ft

19   **Holt Algebra 2**

Practice B
13-3 The Unit Circle

Convert each measure from degrees to radians or from radians to degrees.

1. $\frac{5\pi}{12}$ _____ $75°$

2. $215°$ _____ $\frac{43\pi}{36}$ radians

3. $-\frac{29\pi}{18}$ _____ $-290°$

4. $-180°$ _____ $-\pi$ radians

5. $\frac{5\pi}{3}$ _____ $300°$

6. $-\frac{7\pi}{6}$ _____ $210°$

7. $400°$ _____ $\frac{20\pi}{9}$ radians

8. $\frac{3\pi}{10}$ _____ $54°$

9. $35°$ _____ $\frac{7\pi}{36}$ radians

Use the unit circle to find the exact value of each trigonometric function.

10. $\cos\frac{2\pi}{3}$ _____ $-\frac{1}{2}$

11. $\tan\frac{5\pi}{4}$ _____ 1

12. $\tan\frac{5\pi}{6}$ _____ $-\frac{\sqrt{3}}{3}$

13. $\sin 315°$ _____ $-\frac{\sqrt{2}}{2}$

14. $\cos 225°$ _____ $-\frac{\sqrt{2}}{2}$

15. $\tan 60°$ _____ $\sqrt{3}$

Use a reference angle to find the exact value of the sine, cosine, and tangent of each angle.

16. $150°$ _____ $\frac{1}{2}; -\frac{\sqrt{3}}{2}; -\frac{\sqrt{3}}{3}$

17. $-225°$ _____ $\frac{\sqrt{2}}{2}; -\frac{\sqrt{2}}{2}; -1$

18. $-300°$ _____ $\frac{\sqrt{3}}{2}; \frac{1}{2}; \sqrt{3}$

19. $\frac{11\pi}{6}$ _____ $-\frac{1}{2}; \frac{\sqrt{3}}{2}; -\frac{\sqrt{3}}{3}$

20. $-\frac{2\pi}{3}$ _____ $-\frac{\sqrt{3}}{2}; -\frac{1}{2}; \sqrt{3}$

21. $\frac{5\pi}{4}$ _____ $-\frac{\sqrt{2}}{2}; -\frac{\sqrt{2}}{2}; 1$

Solve.

22. San Antonio, Texas, is located about 30° north of the equator. If Earth's radius is about 3959 miles, approximately how many miles is San Antonio from the equator?

_____ 2073 mi

20   **Holt Algebra 2**

Practice C
13-3 The Unit Circle

Convert each measure from degrees to radians or from radians to degrees.

1. $-\frac{3\pi}{2}$ _____ $-270°$

2. $450°$ _____ $\frac{5\pi}{2}$ radians

3. $\frac{5\pi}{18}$ _____ $50°$

4. $-200°$ _____ $-\frac{10\pi}{9}$ radians

5. $\frac{7\pi}{4}$ _____ $315°$

6. $-\frac{11\pi}{6}$ _____ $-330°$

7. $350°$ _____ $\frac{35\pi}{18}$ radians

8. $\frac{7\pi}{20}$ _____ $63°$

9. $12°$ _____ $\frac{\pi}{15}$ radians

10. $\frac{13\pi}{10}$ _____ $234°$

11. $222°$ _____ $\frac{37\pi}{30}$ radians

12. $-105°$ _____ $-\frac{7\pi}{12}$ radians

Find the exact value of the sine, cosine, and tangent of each angle.

13. $330°$ _____ $-\frac{1}{2}; \frac{\sqrt{3}}{2}; -\frac{\sqrt{3}}{3}$

14. $\frac{7\pi}{4}$ _____ $-\frac{\sqrt{2}}{2}; \frac{\sqrt{2}}{2}; -1$

15. $240°$ _____ $-\frac{\sqrt{3}}{2}; -\frac{1}{2}; \sqrt{3}$

16. $\frac{5\pi}{6}$ _____ $\frac{1}{2}; -\frac{\sqrt{3}}{2}; -\frac{\sqrt{3}}{3}$

17. $225°$ _____ $-\frac{\sqrt{2}}{2}; -\frac{\sqrt{2}}{2}; 1$

18. $120°$ _____ $\frac{\sqrt{3}}{2}; -\frac{1}{2}; -\sqrt{3}$

19. $45°$ _____ $\frac{\sqrt{2}}{2}; \frac{\sqrt{2}}{2}; 1$

20. $-\pi$ _____ $0; -1; 0$

21. $-\frac{5\pi}{6}$ _____ $-\frac{1}{2}; -\frac{\sqrt{3}}{2}; \frac{\sqrt{3}}{3}$

22. $-\frac{\pi}{4}$ _____ $-\frac{\sqrt{2}}{2}; \frac{\sqrt{2}}{2}; -1$

23. $-\frac{\pi}{3}$ _____ $-\frac{\sqrt{3}}{2}; \frac{1}{2}; -\sqrt{3}$

24. $135°$ _____ $\frac{\sqrt{2}}{2}; -\frac{\sqrt{2}}{2}; -1$

Solve.

25. A pendulum is 18 feet long. Its central angle is 44°. The pendulum makes one back and forth swing every 12 seconds. To the nearest foot, how far does the pendulum swing each minute?

_____ 138 ft

21   **Holt Algebra 2**

Reteach
13-3 The Unit Circle

Radians are a real number measure of rotation.
To convert between radians and degrees, use the following identity.

$$\pi \text{ radians} = 180°$$

To convert from degrees to radians, solve the identity for 1 degree.	To convert from radians to degrees, solve the identity for 1 radian.
$1 \text{ degree} = \frac{\pi \text{ radians}}{180°}$	$1 \text{ radian} = \frac{180°}{\pi \text{ radians}}$

Convert 60° to radians.

$$60° = 60°\left(\frac{\pi \text{ radians}}{180°\,^3}\right) = \frac{\pi}{3} \text{ radians}$$

Use dimensional analysis to help. Notice that the degrees cancel so the remaining unit is radians.

Convert $\frac{5\pi}{4}$ radians to degrees.

$$\frac{5\pi}{4} \text{ radians} = \left(\frac{5\pi}{4} \text{ radians}\right)\left(\frac{\overset{45}{180°}}{\pi \text{ radians}}\right) = 225°$$

The radians cancel so the remaining unit is degrees.

Convert each measure from degrees to radians.

1. $-45°$

$$-45° = -45°\left(\frac{\pi \text{ radians}}{180°}\right)$$

_____ $-\frac{\pi}{4}$ radians

2. $150°$

$$150° = 150°\left(\frac{\pi \text{ radians}}{180°}\right)$$

_____ $\frac{5\pi}{6}$ radians

3. $210°$ _____ $\frac{7\pi}{6}$ radians

4. $-120°$ _____ $-\frac{2\pi}{3}$ radians

Convert each measure from radians to degrees.

5. $\frac{4\pi}{3}$ radians

$$\frac{4\pi}{3} \text{ radians} = \left(\frac{4\pi}{3} \text{ radians}\right)\left(\frac{180°}{\pi \text{ radians}}\right)$$

_____ $240°$

6. $-\frac{3\pi}{2}$ radians

$$-\frac{3\pi}{2} \text{ radians} = \left(-\frac{3\pi}{2} \text{ radians}\right)\left(\frac{180°}{\pi \text{ radians}}\right)$$

_____ $-270°$

7. $\frac{\pi}{6}$ radians _____ $30°$

8. $\frac{5\pi}{3}$ radians _____ $300°$

22   **Holt Algebra 2**

Holt Algebra 2

Reteach
The Unit Circle (continued)

Use a reference angle to find the exact value of the sine, cosine, and tangent of an angle in any quadrant.

Find the value of the trigonometric functions of 150°.

Step 1 Sketch the angle.
Find the measure of the reference angle.
$180° - 150° = 30°$

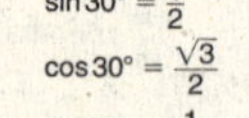

Step 2 Draw the triangle with the reference angle.
Label the sides with their lengths.

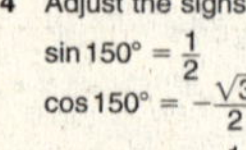

Step 3 Find the sine, cosine, and tangent of 30°.
$$\sin 30° = \frac{1}{2}$$
$$\cos 30° = \frac{\sqrt{3}}{2}$$
$$\tan 30° = \frac{1}{\sqrt{3}}$$

Step 4 Adjust the signs for 150°.
$$\sin 150° = \frac{1}{2}$$
$$\cos 150° = -\frac{\sqrt{3}}{2}$$
$$\tan 150° = -\frac{1}{\sqrt{3}}$$

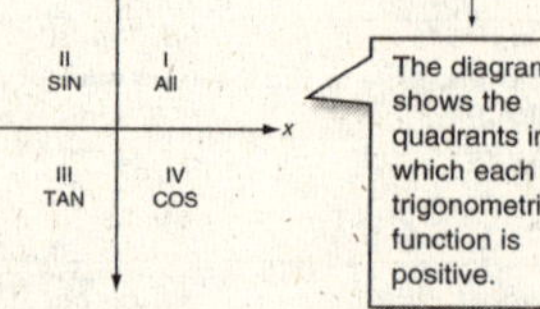

The diagram shows the quadrants in which each trigonometric function is positive.

Complete to find the exact value of the sine, cosine, and tangent of 315°.

9. Find the measure of the reference angle. **45°**

10. Find the sine, cosine, and tangent of the reference angle.
$$\sin 45° = \frac{\sqrt{2}}{2}$$
$$\cos 45° = \frac{\sqrt{2}}{2}$$
$$\tan 45° = 1$$

11. Adjust the signs to find the sine, cosine, and tangent of 315°.
$$\sin 315° = -\frac{\sqrt{2}}{2}$$
$$\cos 315° = \frac{\sqrt{2}}{2}$$
$$\tan 315° = -1$$

Challenge
A Radian as a Measure of Length

Radians can be useful in application problems because of the close connection between radian measure for angles and arc length in a circle. Consider the formula for finding the length of an arc: $s = r\theta$. The variables s and r are both lengths, s is the length of the arc and r is the radius of the circle, and θ is an angle measured in radians. A radian is sometimes considered a unitless measure since it measures the distance around a circle using the radius as the measure of length. For instance, an angle of 2 radians is related to an arc of 2 radii around the circle. It makes no difference the size of the circle; an arc of 2 radii around the circle still produces the same angle.

Solve the following application problems by using the formula $s = r\theta$.

1. A captain at sea measures the distance his ship travels in nautical miles. A nautical mile is the length of the arc along the surface of Earth that is intercepted by an angle of 1 minute $\left(\frac{1}{60}\right)°$. If the radius of Earth is 3960 miles (1 land mile is 5280 feet), find the length of a nautical mile to the nearest 10 feet.

6080 ft

2. One year has approximately 365 days. Earth travels around the sun in an approximately circular orbit with a radius of 93 million miles. Find the distance Earth travels in its orbit each day. Calculate the speed of Earth in its orbit in miles per hour. Round answers to the nearest whole number.

1,600,921 mi; 66,705 mi/h

3. Use the formula for the area of a circle, $A = \pi r^2$, and derive the formula for the area of a sector using only the variables r and θ. Assume that θ is in radians.

Area of circle $= \pi r^2$; A sector whose central angle has a measure of θ radians has an area of $\frac{\theta}{2\pi}$ times the area of the circle.

So Area of sector $= \frac{\theta}{2\pi}(\pi r^2) = \frac{1}{2}\theta r^2$.

4. In the diagram below, find the area of the shaded sector. The circles are tangent to one another as shown in the diagram. **$\frac{\pi}{4}$**

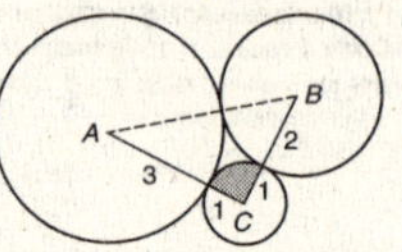

Problem Solving
The Unit Circle

Gabe is spending two weeks on an archaeological dig. He finds a fragment of a circular plate that his leader thinks may be valuable. The arc length of the fragment is about $\frac{1}{6}$ the circumference of the original complete plate and measures 1.65 inches.

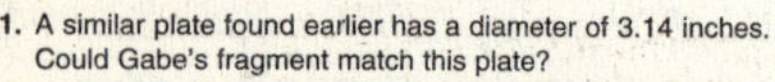

1. A similar plate found earlier has a diameter of 3.14 inches. Could Gabe's fragment match this plate?

   a. Write an expression for the radius, r, of the earlier plate. $r = \frac{\pi}{2}$

   b. What is the measure, in radians, of a central angle, θ, that intercepts an arc that is $\frac{1}{6}$ the length of the circumference of a circle? $\theta = \frac{2\pi}{6}$ or $\frac{\pi}{3}$

   c. Write an expression for the arc length, S, intercepted by this central angle. $S = r\theta = \frac{\pi}{2} \cdot \frac{\pi}{3} = \frac{\pi^2}{6}$

   d. How long would the arc length of a fragment be if it were $\frac{1}{6}$ the circumference of the plate? **1.64 in.**

   e. Could Gabe's plate be a matching plate? Explain.
   Yes; possible answer: because the arc length of the fragment is very close to the arc length that would be expected for a plate of diameter π

2. Toby finds another fragment of arc length 2.48 inches. What fraction of the outer edge of Gabe's plate would it be if this fragment were part of Gabe's plate? $\frac{1}{4}$

The diameter of a merry-go-round at the playground is 12 feet. Elijah stands on the edge and his sister pushes him around. Choose the letter for the best answer.

3. How far does Elijah travel if he moves through an angle of $\frac{5\pi}{4}$ radians?

   A 12.0 ft   C) 23.6 ft
   B 15.1 ft   D 47.1 ft

4. Through what angle does Elijah move if he travels a distance of 80 feet around the circumference?

   F $\frac{40}{3}\pi$ radians   H) $\frac{40}{3}$ radians
   G $\frac{80}{3}$ radians   J $\frac{20}{3}$ radians

Virgil sets his boat on a 1000-yard course keeping a constant distance from a rocky outcrop. Choose the letter for the best answer.

5. If Virgil keeps a distance of 200 yards, through what angle does he travel?

   A 5π radians   C 10 radians
   B) 5 radians   D 10π radians

6. If Virgil keeps a distance of 500 yards, what fraction of the circumference of a circle does he cover?

   F) $\frac{1}{\pi}$ H $\frac{3}{4\pi}$
   G $\frac{\pi}{3}$ J $\frac{3\pi}{4}$

Reading Strategy
Use a Visual Map

The unit circle shown at right has a radius of 1 unit. So the terminal side of each angle has a length of 1 unit. As shown in the diagram below, that corresponds to the hypotenuse of a right triangle. The ordered pair shows the length of each side of the triangle. Notice the ordered pairs for the points at 0°, 90°, 180°, and 270°.

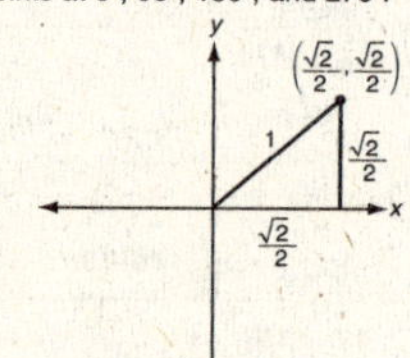

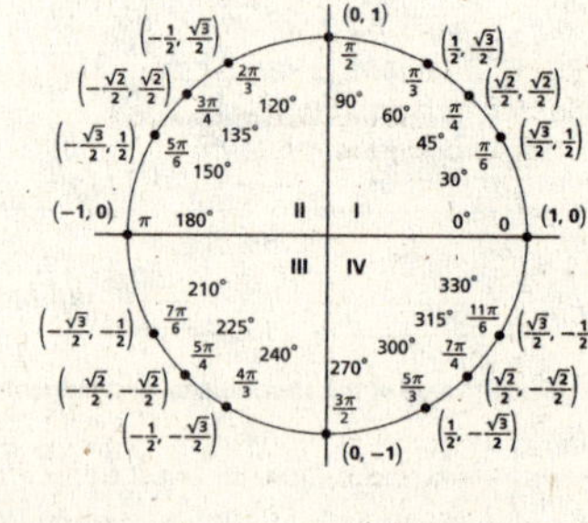

On the circle, for every value of θ,
$$\sin\theta = \frac{y}{r} = \frac{y}{1} = y$$
$$\cos\theta = \frac{x}{r} = \frac{x}{1} = x$$
$$\tan\theta = \frac{y}{x}$$

Use the unit circle to answer each question.

1. Express 360° in radians. **2π**

2. What is sin360°? **0**

3. What is cos360°? **1**

4. a. In which quadrant does the terminal side of a 150° angle lie? **Quadrant II**

   b. Express 150° in radians. **$\frac{5\pi}{6}$**

   c. What point does the terminal side of an angle of 150° pass through on the unit circle? **$\left(-\frac{\sqrt{3}}{2}, \frac{1}{2}\right)$**

   d. Which coordinate of the ordered pair represents sin150°? **$\frac{1}{2}$**

   e. Which coordinate of the ordered pair represents cos150°? **$-\sqrt{\frac{3}{2}}$**

   f. Write an expression for tan150° and simplify. **$\dfrac{\frac{1}{2}}{-\frac{\sqrt{3}}{2}} = -\frac{\sqrt{3}}{3}$**

Practice A
Inverses of Trigonometric Functions

Find all possible values of each expression.

1. $\cos^{-1}\left(-\frac{\sqrt{3}}{2}\right)$

   a. Since $\cos\theta = x$, look at the x-coordinates on the unit circle. Find the points where the x-coordinate is $-\frac{\sqrt{3}}{2}$. $\quad \frac{5\pi}{6}, \frac{7\pi}{6}$

   b. Add integer multiples of $2\pi n$. $\quad \frac{5\pi}{6} + 2\pi n; \frac{7\pi}{6} + 2\pi n$

2. $\sin^{-1}\left(\frac{1}{2}\right)$ $\quad \frac{\pi}{6} + 2\pi n; \frac{5\pi}{6} + 2\pi n$

3. $\cos^{-1}\left(\frac{1}{2}\right)$ $\quad \frac{\pi}{3} + 2\pi n; \frac{5\pi}{3} + 2\pi n$

4. $\tan^{-1}1$ $\quad \frac{\pi}{4} + 2\pi n; \frac{5\pi}{4} + 2\pi n$

5. $\sin^{-1}\left(\frac{\sqrt{3}}{2}\right)$ $\quad \frac{\pi}{3} + 2\pi n; \frac{2\pi}{3} + 2\pi n$

6. $\cos^{-1}\left(-\frac{1}{2}\right)$ $\quad \frac{2\pi}{3} + 2\pi n; \frac{4\pi}{3} + 2\pi n$

Evaluate each inverse trigonometric function. Give your answer in both radians and degrees.

7. $\text{Sin}^{-1}0$

   a. The inverse sine function is restricted to which quadrants? $\quad$ I and IV

   b. Find the value of θ whose sine is 0. $\quad$ 0 radians, 0°

8. $\text{Sin}^{-1}\left(-\frac{\sqrt{2}}{2}\right)$ $\quad \frac{7\pi}{4}; 315°$

9. $\text{Tan}^{-1}(\sqrt{3})$ $\quad \frac{\pi}{3}; 60°$

10. $\text{Cos}^{-1}\left(\frac{\sqrt{3}}{2}\right)$ $\quad \frac{\pi}{6}; 30°$

11. $\text{Tan}^{-1}(-1)$ $\quad \frac{7\pi}{4}; 315°$

12. $\text{Cos}^{-1}(-1)$ $\quad \pi; 180°$

13. $\text{Sin}^{-1}(-1)$ $\quad \frac{3\pi}{2}; 270°$

Solve each equation to the nearest tenth. Use the given restrictions.

14. $\sin\theta = 0.95$, for $-90° \leq \theta \leq 90°$ $\quad$ 71.8°

15. $\cos\theta = 0.2$, for $270° \leq \theta \leq 360°$ $\quad$ 281.5°

Solve.

16. The pilot of a plane wants to fly 800 miles east and 155 miles north of the airport. To the nearest degree, in what direction should he head? $\quad$ 11° north of east

Holt Algebra 2

Practice B
Inverses of Trigonometric Functions

Find all possible values of each expression.

1. $\sin^{-1}\left(-\frac{\sqrt{3}}{2}\right)$ $\quad \frac{4\pi}{3} + 2\pi n; \frac{5\pi}{3} + 2\pi n$

2. $\cos^{-1}\left(-\frac{1}{2}\right)$ $\quad \frac{2\pi}{3} + 2\pi n; \frac{4\pi}{3} + 2\pi n$

3. $\tan^{-1}0$ $\quad 0 + 2\pi n; \pi + 2\pi n$

4. $\sin^{-1}\left(-\frac{\sqrt{2}}{2}\right)$ $\quad \frac{5\pi}{4} + 2\pi n; \frac{7\pi}{4} + 2\pi n$

5. $\cos^{-1}\left(-\frac{\sqrt{2}}{2}\right)$ $\quad \frac{3\pi}{4} + 2\pi n; \frac{5\pi}{4} + 2\pi n$

6. $\tan^{-1}\left(\frac{\sqrt{3}}{3}\right)$ $\quad \frac{\pi}{6} + 2\pi n; \frac{7\pi}{6} + 2\pi n$

Evaluate each inverse trigonometric function. Give your answer in both radians and degrees.

7. $\text{Sin}^{-1}(-1)$ $\quad \frac{3\pi}{2}; 270°$

8. $\text{Tan}^{-1}(-\sqrt{3})$ $\quad \frac{5\pi}{3}; 300°$

9. $\text{Cos}^{-1}1$ $\quad 0; 0°$

10. $\text{Sin}^{-1}\left(\frac{\sqrt{3}}{2}\right)$ $\quad \frac{\pi}{3}; 60°$

11. $\text{Tan}^{-1}\left(-\frac{\sqrt{3}}{3}\right)$ $\quad \frac{11\pi}{6}; 330°$

12. $\text{Cos}^{-1}\left(\frac{\sqrt{2}}{2}\right)$ $\quad \frac{\pi}{4}; 45°$

Solve each equation to the nearest tenth. Use the given restrictions.

13. $\sin\theta = 0.45$, for $0° < \theta < 90°$ $\quad$ 26.7°

14. $\sin\theta = 0.801$, for $90° < \theta < 270°$ $\quad$ 233.2°

15. $\tan\theta = 2.42$, for $180° < \theta < 360°$ $\quad$ 247.5°

16. $\cos\theta = -0.334$, for $0° < \theta < 180°$ $\quad$ 109.5°

17. $\cos\theta = -0.181$, for $180° < \theta < 360°$ $\quad$ 259.6°

18. $\tan\theta = -10$, for $90° < \theta < 270°$ $\quad$ 95.7°

Solve.

19. A 21-foot ladder is leaning against a building. The base of the ladder is 7 feet from the base of a building. To the nearest degree, what is the measure of the angle that the ladder makes with the ground? $\quad$ 71°

Holt Algebra 2

Practice C
Inverses of Trigonometric Functions

Find all possible values of each expression.

1. $\sin^{-1}\left(\frac{\sqrt{3}}{2}\right)$ $\quad \frac{\pi}{3} + 2\pi n; \frac{2\pi}{3} + 2\pi n$

2. $\cos^{-1}\left(\frac{\sqrt{2}}{2}\right)$ $\quad \frac{\pi}{4} + 2\pi n; \frac{7\pi}{4} + 2\pi n$

3. $\tan^{-1}(-\sqrt{3})$ $\quad \frac{2\pi}{3} + 2\pi n; \frac{5\pi}{3} + 2\pi n$

4. $\cos^{-1}\left(-\frac{\sqrt{2}}{2}\right)$ $\quad \frac{3\pi}{4} + 2\pi n; \frac{5\pi}{4} + 2\pi n$

5. $\sin^{-1}\left(\frac{1}{2}\right)$ $\quad \frac{\pi}{6} + 2\pi n; \frac{5\pi}{6} + 2\pi n$

6. $\tan^{-1}(-1)$ $\quad \frac{3\pi}{4} + 2\pi n; \frac{7\pi}{4} + 2\pi n$

Evaluate each inverse trigonometric function. Give your answer in both radians and degrees.

7. $\text{Sin}^{-1}\left(-\frac{\sqrt{3}}{2}\right)$ $\quad \frac{5\pi}{3}; 300°$

8. $\text{Cos}^{-1}\left(-\frac{1}{2}\right)$ $\quad \frac{2\pi}{3}; 120°$

9. $\text{Tan}^{-1}\left(\frac{\sqrt{3}}{3}\right)$ $\quad \frac{\pi}{6}; 30°$

10. $\text{Sin}^{-1}(1)$ $\quad \frac{\pi}{2}; 90°$

11. $\text{Tan}^{-1}(\sqrt{3})$ $\quad \frac{\pi}{3}; 60°$

12. $\text{Cos}^{-1}0$ $\quad \frac{\pi}{2}; 90°$

Solve each equation to the nearest tenth. Use the given restrictions.

13. $\sin\theta = -0.204$, for $90° < \theta < 270°$ $\quad$ 191.8°

14. $\tan\theta = 8.055$, for $90° < \theta < 270°$ $\quad$ 262.9°

15. $\cos\theta = 0.778$, for $180° < \theta < 360°$ $\quad$ 321.1°

16. $\tan\theta = -4$, for $0° < \theta < 180°$ $\quad$ 104.0°

17. $\sin\theta = 0.094$, for $90° < \theta < 270°$ $\quad$ 174.6°

18. $\cos\theta = -0.555$, for $90° < \theta < 270°$ $\quad$ 123.7°; 236.3°

Solve.

19. A gutter at the edge of a roof drops 2 inches for every 30 feet of length. To the nearest tenth of a degree, what is the measure of the angle that the gutter makes with the roof? $\quad$ 0.3°

Holt Algebra 2

Reteach
Inverses of Trigonometric Functions

The trigonometric functions have inverse relations.

Trigonometric Function	Inverse Relation
$\sin\theta = a$	$\sin^{-1}a = \theta$
$\cos\theta = a$	$\cos^{-1}a = \theta$
$\tan\theta = a$	$\tan^{-1}a = \theta$

Read "the inverse sine of a."
Read "the inverse cosine of a."
Read "the inverse tangent of a."

Evaluate the inverse of a trigonometric function.

Step 1 Each inverse trigonometric relation has multiple values. Think, "What reference angle has the given trigonometric value?" Find the two angles between 0 and 2π radians that have the given value for the trigonometric function.

Step 2 Add $(2\pi)n$ to each angle to represent all the coterminal angles, where n is an integer.

Find all possible values of $\cos^{-1}\frac{\sqrt{3}}{2}$.

Think: What reference angle has a cosine of $\frac{\sqrt{3}}{2}$?

Step 1 The value of the cosine is positive.
Cosine is positive in Quadrants I and IV.
In Quadrant I, the reference angle is the angle.
$\cos\frac{\pi}{6} = \frac{\sqrt{3}}{2}$ so $\cos^{-1}\frac{\sqrt{3}}{2} = \frac{\pi}{6}$
To find the angle in Quadrant IV,
subtract the reference angle from 2π.
$2\pi - \frac{\pi}{6} = \frac{11\pi}{6}$

Step 2 Add $(2\pi)n$ to each angle.
$\frac{\pi}{6} + 2\pi n$ or $\frac{11\pi}{6} + 2\pi n$

Complete to find all possible values of $\tan^{-1}\sqrt{3}$.

1. In which quadrants is tangent positive? $\quad$ Quadrants I, III

2. Name one angle that has tangent of $\sqrt{3}$. $\quad \frac{\pi}{3}$

3. Name another angle that has tangent of $\sqrt{3}$. $\quad \frac{4\pi}{3}$

4. Add $(2\pi)n$ to each angle to find all possible values. $\quad \frac{\pi}{3} + 2\pi n; \frac{4\pi}{3} + 2\pi n$

Holt Algebra 2

Holt Algebra 2

Reteach
Inverses of Trigonometric Functions (continued)

When their domains are restricted, the inverse trigonometric relations can be defined as functions. These inverse trigonometric functions can be distinguished from the inverse trigonometric relations by the capital letters used to denote the functions.

Use the range of the inverse functions to help you evaluate the functions.

- For $\text{Sin}^{-1}a = \theta$, θ lies in Quadrants I and IV and $-\frac{\pi}{2} \le \theta \le \frac{\pi}{2}$.
- For $\text{Cos}^{-1}a = \theta$, θ lies in Quadrants I and II and $0 \le \theta \le \pi$.
- For $\text{Tan}^{-1}a = \theta$, θ lies in Quadrants I and IV and $-\frac{\pi}{2} < \theta < \frac{\pi}{2}$.

Note that for $\text{Sin}^{-1}a$ and $\text{Cos}^{-1}a$, the terminal side of θ can lie on one of the axes.

For example: $\text{Sin}^{-1}(-1) = -\frac{\pi}{2}$ or $-90°$.

Evaluate $\text{Tan}^{-1}\left(-\frac{\sqrt{3}}{3}\right)$.

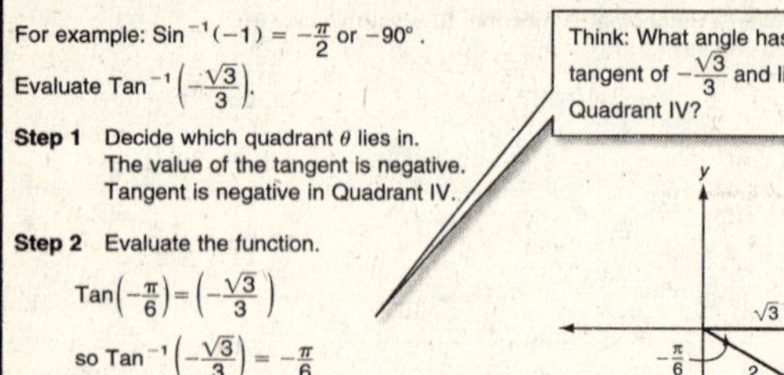

Step 1 Decide which quadrant θ lies in. The value of the tangent is negative. Tangent is negative in Quadrant IV.

Step 2 Evaluate the function.

$$\text{Tan}\left(-\frac{\pi}{6}\right) = \left(-\frac{\sqrt{3}}{3}\right)$$

so $\text{Tan}^{-1}\left(-\frac{\sqrt{3}}{3}\right) = -\frac{\pi}{6}$

Complete to evaluate each trigonometric function.

5. $\text{Cos}^{-1}\frac{\sqrt{2}}{2}$

   a. Is the value of cosine negative or positive? — **Positive**

   b. In which quadrant does θ lie? — **Quadrant I**

   c. What angle in the range has cosine of $\frac{\sqrt{2}}{2}$? — $\frac{\pi}{4}$

6. $\text{Sin}^{-1}\left(-\frac{1}{2}\right)$

   a. Is the value of sine negative or positive? — **Negative**

   b. In which quadrant does θ lie? — **Quadrant IV**

   c. What angle in the range has sine of $-\frac{1}{2}$? — $-\frac{\pi}{6}$

31
Holt Algebra 2

Challenge
Combination Trigonometric Functions

When evaluating an inverse trigonometric function, it is helpful to think of the result as an angle. For example, $\text{Sin}^{-1}x$ can be thought of as an angle whose sine is x. The other common notation for $\text{Sin}^{-1}x$ is arcsine x and is read as an angle whose sine is x.

Thinking of the inverse trigonometric functions in this manner will help you evaluate combinations of trigonometric functions and inverse trigonometric functions. For example, consider this expression.

$$\sin\left(\text{Cos}^{-1}\frac{3}{5}\right)$$

The expression directs you to find the sine of an angle θ such that $\theta = \text{Cos}^{-1}\frac{3}{5}$. This is interpreted as θ is an angle whose cosine is $\frac{3}{5}$. Think of θ as an angle in a right triangle as shown at right. The cosine of angle θ is $\frac{3}{5}$ since that is the ratio of the adjacent side to the hypotenuse. The third side of the triangle is 4, and the ratio of the opposite side to the hypotenuse is $\frac{4}{5}$. It follows that $\sin\left(\text{Cos}^{-1}\frac{3}{5}\right) = \frac{4}{5}$.

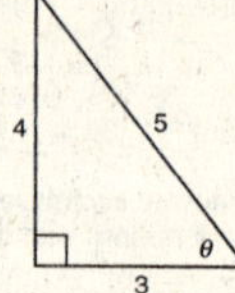

Evaluate each of the following.

1. $\sin\left(\text{Cos}^{-1}\frac{4}{5}\right)$ — $\frac{3}{5}$

2. $\cos\left(\text{Tan}^{-1}\frac{3}{4}\right)$ — $\frac{4}{5}$

3. $\tan\left(\text{Sin}^{-1}\frac{5}{13}\right)$ — $\frac{5}{12}$

4. $\cos\left(\text{Sin}^{-1}\frac{12}{13}\right)$ — $\frac{5}{13}$

5. $\sec\left(\text{Tan}^{-1}\frac{7}{24}\right)$ — $\frac{25}{24}$

6. $\csc\left(\text{Sin}^{-1}\frac{9}{41}\right)$ — $\frac{41}{9}$

7. $\cos\left(\text{Sin}^{-1}x\right)$ where $-1 \le x \le 1$
   (*Hint:* Think of x as a ratio and use the Pythagorean Theorem.)

   $$\sqrt{1 - x^2}$$

8. $\csc\left(\text{Tan}^{-1}\frac{1}{x}\right)$ where $x > 0$
   (*Hint:* Draw a triangle and use the Pythagorean Theorem.)

   $$\sqrt{1 + x^2}$$

32
Holt Algebra 2

Problem Solving
Inverses of Trigonometric Functions

Rafe is concerned that some recently constructed buildings in his town do not comply with code restrictions. New buildings are limited to a maximum of 40 feet in height.

1. When working with angles of elevation from his eye level, Rafe realizes that he must allow for his own height, 5 feet 9 inches, in his calculations. Explain how he can do this.
   Possible answer: Subtract his height from any height measurements.

2. On the building plans, the height of the new bank is 33 feet. Rafe calculates what the angle of elevation should be from 100 feet away if the bank is 33 feet tall.

   a. Label the diagram to show the height of the bank that Rafe will use in his calculations Mark the angle of elevation.

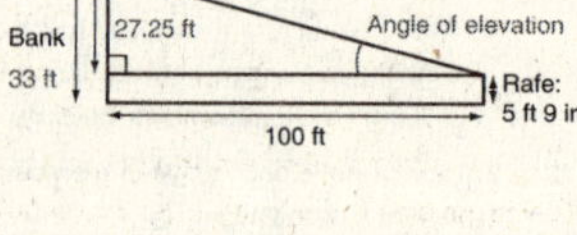

   b. Write a trigonometric function for the assumed angle of elevation, θ

   $$\theta = \tan^{-1}\left(\frac{27.25}{100}\right)$$

   c. To the nearest tenth of a degree, what is the assumed angle of elevation? — **15.2°**

   d. Using a clinometer, Rafe measures the angle of elevation to be 14.2°. Does the bank comply with the building code? Explain.

   Yes; since the actual angle of elevation is less than the assumed angle of elevation, the building is less than 33 ft tall.

3. On the plans, the height of the new inn is 39 feet. Rafe finds the angle of elevation and compares it to what he measures.

   a. Predict the angle of elevation of the highest point on this building from a distance of 100 feet and allowing for Rafe's height. — **18.4°**

   b. Predict the angle of elevation of the highest point on a building that is 40 feet tall, allowing for Rafe's height. — **18.9°**

   c. If Rafe measures an angle of elevation of 19.0°, how does the height of the inn compare to its declared height of 39 feet? Explain.

   The inn is more than 40 ft tall. Since the angle of elevation is greater than 18.9°, then the building is taller than 40 ft.

4. Carrie says the angle of elevation of the top of the flagpole at school is 55.9° from a distance of 20 feet away and allowing for her height of 5 feet 6 inches.

   a. Write and evaluate an expression for the height of the flagpole to the nearest tenth of a foot. — $20\tan 55.9° + 5.5 = 35.0$ ft

   b. What should be the angle of elevation if Rafe measures it from a distance of 50 feet away? Write and evaluate an expression.

   $$\tan^{-1}\left(\frac{35 - 5.75}{50}\right) = 30.3°$$

33
Holt Algebra 2

Reading Strategy
Follow a Procedure

You can find the inverse of trigonometric functions, such as sine and cosine, using a two-step operation.

	EXAMPLE	STEP 1	STEP 2
For the inverse sine function, use the sine inverse relation $\sin^{-1}a = \theta$.	Find all possible values of $\sin^{-1}\frac{1}{2}$.	Using the unit circle, find the values between 0 and 2π for which $\sin\theta = \frac{1}{2}$. $\sin\frac{\pi}{6} = \frac{1}{2}$, $\sin\frac{5\pi}{6} = \frac{1}{2}$	Find angles that are coterminal with the angles $\frac{\pi}{6}$ and $\frac{5\pi}{6}$. $\frac{\pi}{6} + (2\pi)n$, $\frac{5\pi}{6} + (2\pi)n$
For the inverse cosine function, use the cosine inverse relation $\cos^{-1}a = \theta$.	Find all possible values of $\cos^{-1}\frac{1}{2}$.	Using the unit circle, find the values between 0 and 2π for which $\cos\theta = \frac{1}{2}$. $\cos\frac{\pi}{3} = \frac{1}{2}$, $\cos\frac{5\pi}{3} = \frac{1}{2}$	Find angles that are coterminal with the angles $\frac{\pi}{3}$ and $\frac{5\pi}{3}$. $\frac{\pi}{3} + (2\pi)n$, $\frac{5\pi}{3} + (2\pi)n$

Answer each question.

1. Explain the meaning of the inverse trigonometric functions.
   Possible answer: The inverse function is used to find the measure of the angle when the value of the trigonometric function is known.

2. Find all values of θ between 0 and 2π for which $\sin\theta = \frac{\sqrt{3}}{2}$. — $\frac{\pi}{3}, \frac{2\pi}{3}$

3. Find all values of θ between 0 and 2π for which $\sin\theta = \frac{\sqrt{2}}{2}$. — $\frac{\pi}{4}, \frac{3\pi}{4}$

4. Find all values of θ between 0 and 2π for which $\cos\theta = \frac{\sqrt{3}}{2}$. — $\frac{\pi}{6}, \frac{11\pi}{6}$

5. Find all possible values of $\cos^{-1}\frac{\sqrt{2}}{2}$.

   a. Find the values of θ between 0 and 2π for which $\cos\theta = \frac{\sqrt{2}}{2}$. — $\frac{\pi}{4}, \frac{7\pi}{4}$

   b. Find angles that are coterminal with these angle values. — $\frac{\pi}{4} + (2\pi)n, \frac{7\pi}{4} + (2\pi)n$

6. Find all possible values of $\sin^{-1} = 0$.

   a. Find the values of θ between 0 and 2π for which $\sin\theta = 0$. — $0, \pi$

   b. Find angles that are coterminal with these angle values. — $(2\pi)n, \pi + (2\pi)n$

   c. Express your answer to part b in degrees. — $360n, 90 + 360n$

34
Holt Algebra 2

Practice A
The Law of Sines

Find the area of each triangle. Round to the nearest tenth.

1. 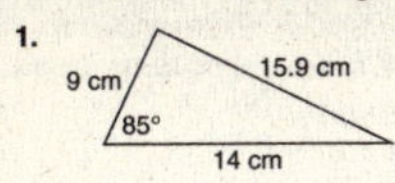

a. Write the formula for the area of a triangle.
$$A = \frac{1}{2}(9)(14)\sin 85$$

b. Substitute the known values and evaluate.
$$62.8 \text{ cm}^2$$

2. 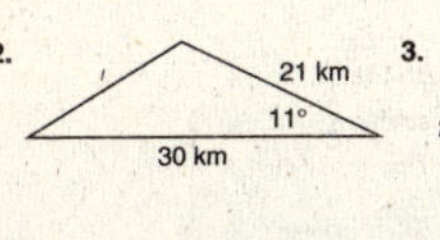
$$60.1 \text{ km}^2$$

3. 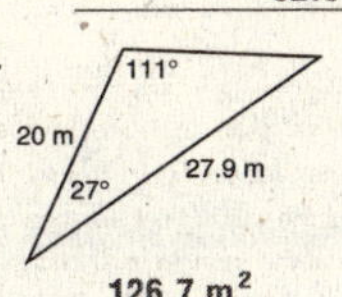
$$126.7 \text{ m}^2$$

4. 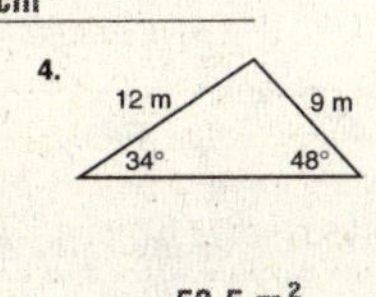
$$53.5 \text{ m}^2$$

Solve each triangle. Round to the nearest tenth.

5. 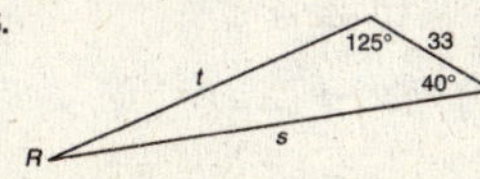

a. Find the measure of the third angle.
$$R = 15°$$

b. Use the Law of Sines to find the unknown side lengths.
$$t \approx 82; \ s \approx 104.4$$

6. 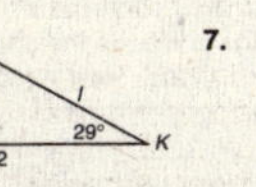
$$m\angle J = 120°;$$
$$k \approx 34.7;$$
$$l \approx 36.9$$

7. 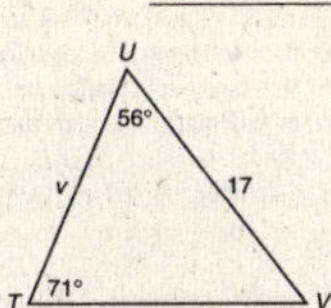
$$m\angle V = 53°;$$
$$v \approx 14.4;$$
$$u \approx 14.9$$

8. 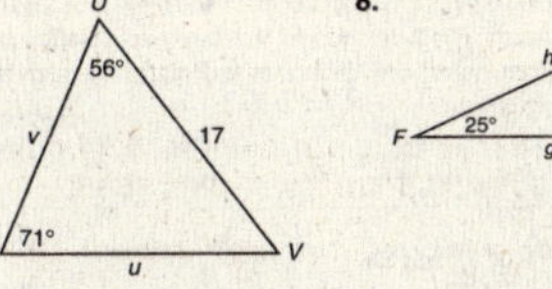
$$m\angle H = 88°;$$
$$h \approx 16.6;$$
$$g \approx 15.2$$

Solve.

9. Two sides of a triangular garden have 6 feet and 9 feet of edging. To the nearest tenth, what is the area of the garden if the angle formed by the edged sides is 80°?
$$26.6 \text{ ft}^2$$

35

Practice B
The Law of Sines

Find the area of each triangle. Round to the nearest tenth.

1. 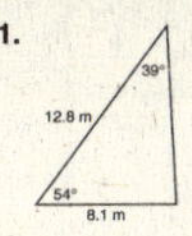
$$41.9 \text{ m}^2$$

2. 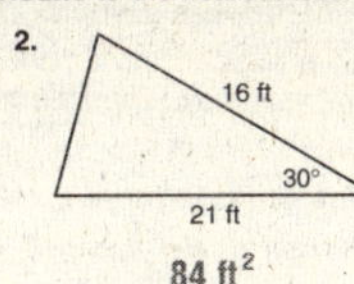
$$84 \text{ ft}^2$$

3. 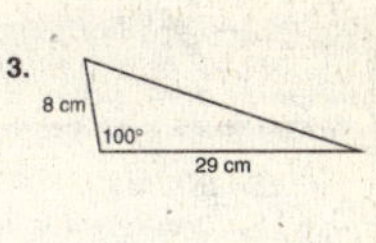
$$114.2 \text{ cm}^2$$

Solve each triangle. Round to the nearest tenth.

4. 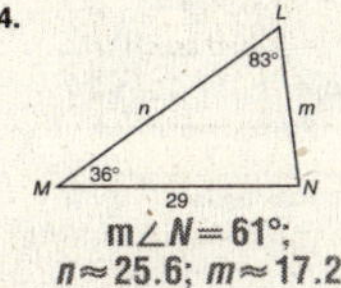
$$m\angle N = 61°;$$
$$n \approx 25.6; \ m \approx 17.2$$

5. 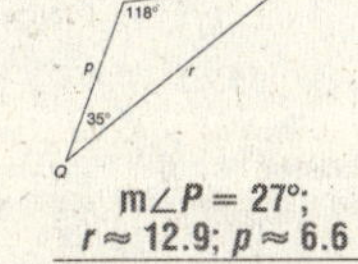
$$m\angle P = 27°;$$
$$r \approx 12.9; \ p \approx 6.6$$

6. 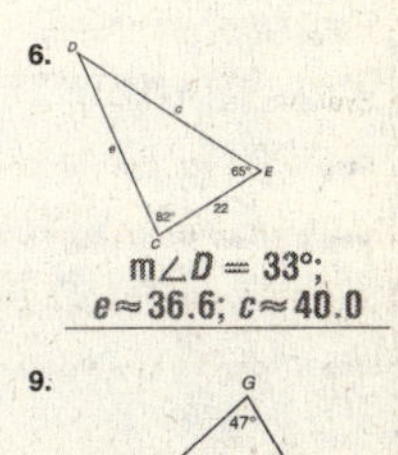
$$m\angle D = 33°;$$
$$e \approx 36.6; \ c \approx 40.0$$

7. 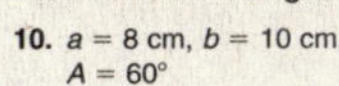
$$m\angle Y = 64°;$$
$$x \approx 10.4; \ y \approx 12.6$$

8.
$$m\angle C = 44°;$$
$$a \approx 48.9; \ b \approx 58.1$$

9.
$$m\angle F = 73°;$$
$$h \approx 7.7; \ f \approx 8.5$$

An artist is designing triangular mosaic tiles. Determine the number of triangles he can form from the given side and angle measures. Then solve the triangles. Round to the nearest tenth.

10. $a = 8$ cm, $b = 10$ cm, $A = 60°$
0 triangles

11. $a = 18$ cm, $b = 15$ cm, $A = 85°$
1 triangle; $c \approx 11.3$ cm; $m\angle B = 56°$; $m\angle C = 39°$

12. $a = 22$ cm, $b = 15$ cm, $A = 120°$
1 triangle; $c \approx 10.3$ cm; $m\angle B = 36°$; $m\angle C = 24°$

Solve.

13. Ann is creating a triangular frame. Two angles and the included side of the frame measure 64°, 58°, and 38 centimeters, respectively. What are the lengths of the other two sides of the frame to the nearest tenth of a centimeter?
38.0 cm; 40.3 cm

36

Practice C
The Law of Sines

Find the area of each triangle. Round to the nearest tenth.

1. 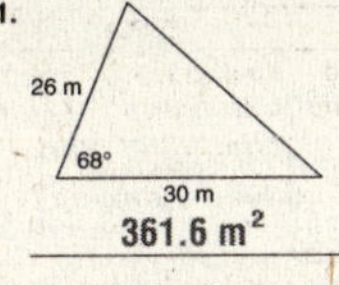
$$361.6 \text{ m}^2$$

2. 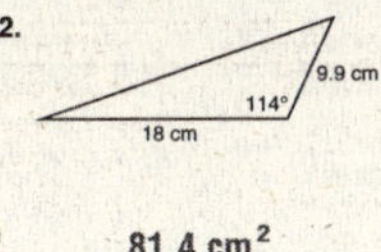
$$81.4 \text{ cm}^2$$

3. 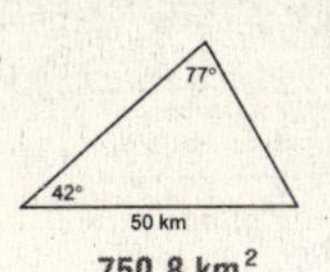
$$750.8 \text{ km}^2$$

Solve each triangle. Round to the nearest tenth.

4. 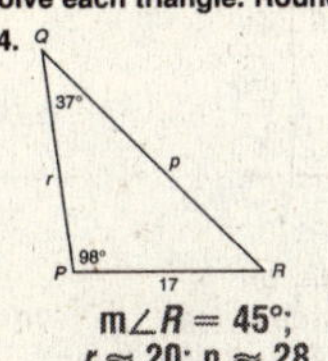
$$m\angle R = 45°;$$
$$r \approx 20; \ p \approx 28$$

5. 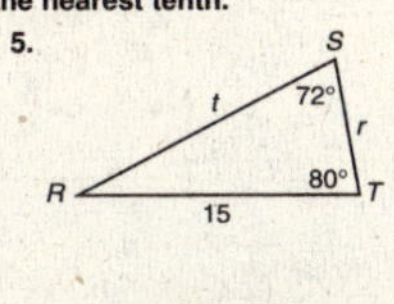
$$m\angle R = 28°;$$
$$t \approx 15.5; \ r \approx 7.4$$

6. 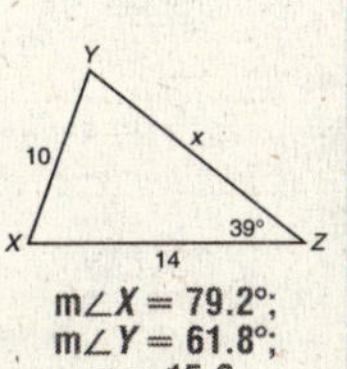
$$m\angle X = 79.2°;$$
$$m\angle Y = 61.8°;$$
$$x \approx 15.6$$

A jeweler is cutting stones into triangles. Determine the number of triangles she can form from the given side and angle measures. Then solve the triangles. Round to the nearest tenth.

7. $a = 8$ cm, $b = 42$ cm, $A = 12°$
0 triangles

8. $a = 21$ cm, $b = 6$ cm, $A = 90°$
1 triangle; $c \approx 20.1$ cm; $m\angle B = 16.6°$; $m\angle C = 73.4°$

9. $a = 7$ cm, $b = 7$ cm, $A = 90°$
0 triangles

10. $a = 27$ cm, $b = 30$ cm, $A = 55°$
2 triangles; $c \approx 28.4$ cm, $m\angle B = 65.5°$, $m\angle C = 59.5°$; $c \approx 6.0$ cm, $m\angle B = 114.5°$, $m\angle C = 10.5°$

11. $a = 29$ cm, $b = 22$ cm, $A = 75°$
1 triangle; $c \approx 25.4$ cm; $m\angle B = 47.1°$; $m\angle C = 57.9°$

12. $a = 4.6$ cm, $b = 9.2$ cm, $A = 30°$
1 triangle; $c \approx 8.0$ cm; $m\angle B = 90°$; $m\angle C = 60°$

Solve.

13. Margaret has two lengths of fence, 20 meters and 24 meters, for two sides of a triangular chicken pen. The third side will be on the north side of a barn. One fence length forms a 75° angle with the barn. How many different pens can she build if one fence is attached at the corner of the barn? What are all the possible lengths for the barn side of the pen?
1 pen; 19.4 m

37

Reteach
The Law of Sines

If you know the lengths of two sides of a triangle and the measure of the angle between the two sides, you can find the area of the triangle.

The angle must be formed by the two sides.

Area of a Triangle $= \frac{1}{2}bc\sin A$

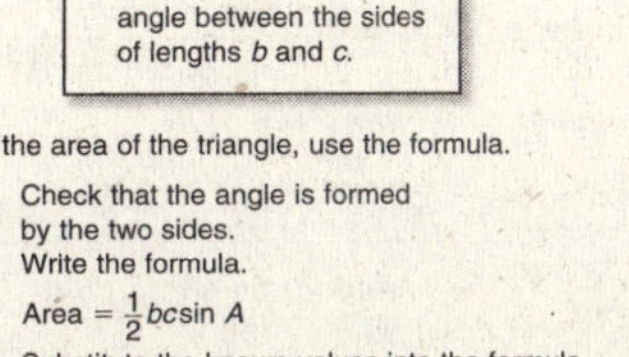

A is the measure of the angle between the sides of lengths b and c.

To find the area of the triangle, use the formula.

Step 1 Check that the angle is formed by the two sides. Write the formula.
$$\text{Area} = \frac{1}{2}bc\sin A$$

Step 2 Substitute the known values into the formula.
$$A = 42°, b = 9, c = 14$$
$$\text{Area} = \frac{1}{2}(9)(14)\sin 42°$$

Check that your calculator is set to degrees to compute the sine.

Step 3 Use a calculator to evaluate the area.
$$\text{Area} \approx 42.1552282$$

Step 4 Round the answer to the nearest tenth. Record the units.
The area is about 42.2 ft².

Complete to find the area of each triangle. Round to the nearest tenth.

1. 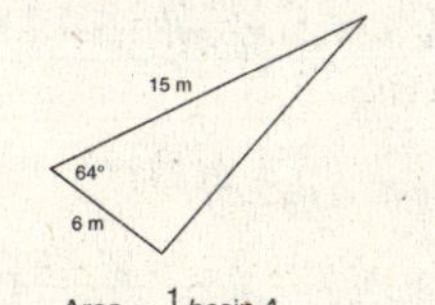

$$\text{Area} = \frac{1}{2}bc\sin A$$
$$A = \underline{64°} \quad b = \underline{6} \quad c = \underline{15}$$
$$40.4 \text{ m}^2$$

2. 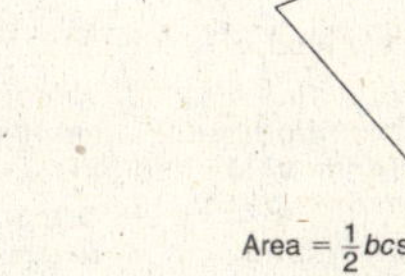

$$\text{Area} = \frac{1}{2}bc\sin A$$
$$A = \underline{76°} \quad b = \underline{16} \quad c = \underline{28}$$
$$217.3 \text{ yd}^2$$

38

59

Reteach
The Law of Sines (continued)

Use the **Law of Sines** to solve a triangle if you know two angle measures and any side length **or** two side lengths and an angle measure that is not between them.

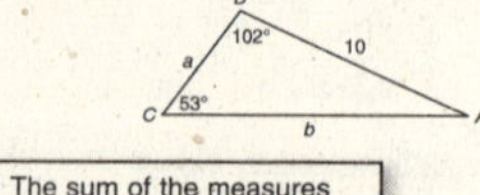

Law of Sines

For $\triangle ABC$: $\dfrac{\sin A}{a} = \dfrac{\sin B}{b} = \dfrac{\sin C}{c}$

Solve the triangle at right.
Round to the nearest tenth.

Step 1 Decide what you know.
$m\angle B = 102°$
$m\angle C = 53°$
$c = 10$

Step 2 Find the missing angle measure.
$53° + 102° + m\angle A = 180°$
$m\angle A = 25°$

Step 3 Use the Law of Sines to find the unknown side lengths.

$\dfrac{\sin A}{a} = \dfrac{\sin C}{c}$ $\dfrac{\sin B}{b} = \dfrac{\sin C}{c}$

$\dfrac{\sin 25°}{a} = \dfrac{\sin 53°}{10}$ $\dfrac{\sin 102°}{b} = \dfrac{\sin 53°}{10}$

$a\sin 53° = 10\sin 25°$ $b\sin 53° = 10\sin 102°$

$a = \dfrac{10\sin 25°}{\sin 53°}$ $b = \dfrac{10\sin 102°}{\sin 53°}$

$a \approx 5.3$ $b \approx 12.2$

Complete to solve the triangle.

3. $m\angle A = \underline{34°}$ $m\angle C = \underline{67°}$ $a = \underline{12}$

4. Find $m\angle B$.

$\underline{79°}$

5. Find b to the nearest tenth.

$\dfrac{\sin A}{a} = \dfrac{\sin B}{b}$; $\dfrac{\sin 34°}{12} = \dfrac{\sin 79°}{b}$ $b \approx 21.1$

6. Find c to the nearest tenth.

$\dfrac{\sin A}{a} = \dfrac{\sin C}{c}$ $c \approx 19.8$

39   **Holt Algebra 2**

Challenge
A Geometric Law of Sines

The Law of Sines can be derived by a geometric argument. Consider the diagram below. The measure of $\angle ACB$ is $\frac{1}{2}$ the measure of $\angle AOB$. Right triangles AOD and BOD are congruent since they share a common leg (side OD) and each hypotenuse is a radius of the circle. So the $m\angle AOD = m\angle BOD = m\angle ACB$. Also $\overline{AD} = \overline{BD} = \frac{1}{2}c$.

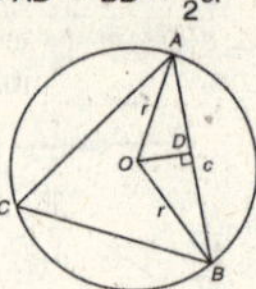

The following equation can be derived from the information given above.

$$\sin \angle AOD = \dfrac{AD}{AO} = \dfrac{\frac{c}{2}}{r} = \sin \angle ACB = \sin \angle C$$

This simplifies to $2r = \dfrac{c}{\sin \angle C}$, which tells us that that the ratio of the length of a side of a triangle to the sine of the opposite angle is equal to twice the radius of the circle circumscribed about the triangle. Thus we obtain the following, which leads immediately to the Law of Sines.

$$2r = \dfrac{a}{\sin \angle A} = \dfrac{b}{\sin \angle B} = \dfrac{c}{\sin \angle C}$$

Solve.

1. A buoy is anchored offshore to mark a sandbar. The straight shoreline at that location runs north and south. From two observation points on the shore 2.4 miles apart, the bearings to the buoy are S46°E and N22°E. How far is the buoy from the shore? What is the distance from the buoy to each of the observation points?
   about 0.70 mile from shore; about 0.97 mile from one point, about 1.86 miles from the second point.

2. The formula for the area of a triangle is Area $= \frac{1}{2}bh$, where b is the length of the base and h is the height of the triangle. Show that Area $= \dfrac{a^2\sin B \sin C}{2\sin A}$.
   In $\triangle ABC$, use $\overline{AC}$, which has length b, as the base. Draw a perpendicular segment from vertex B to line AC. The length h is the height of $\triangle ABC$ and $h = a\sin C$. By the Law of Sines, $\dfrac{\sin A}{a} = \dfrac{\sin B}{b}$, and $b = \dfrac{a\sin B}{\sin A}$.

   Substitute in the area formula: Area $= \dfrac{1}{2}\left(\dfrac{a\sin B}{\sin A}\right)(a\sin C) = \dfrac{a^2\sin B \sin C}{2\sin A}$.

40   **Holt Algebra 2**

Problem Solving
The Law of Sines

In the middle of town, State and Elm streets meet at an angle of 40°. A triangular pocket park between the streets stretches 100 yards along State Street and 53.2 yards along Elm Street. Hoa and Cat walk around the pocket park every day at lunchtime.

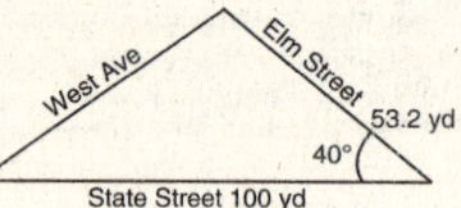

1. Hoa would like to know the area of the pocket park.
   a. Write a formula for the area of the pocket park using the given dimensions.
      $A = \dfrac{1}{2}(100)(53.2)\sin(40°)$
   b. What is the area of the pocket park to the nearest tenth of a square yard?
      1709.8 yd^2

2. Cat determines that the total distance around the pocket park is 221.6 yards.
   a. Write and evaluate an expression for the length of the park along West Avenue.
      $221.6 - (100 + 53.2) = 68.4$ **yd**
   b. Use the Law of Sines to find $\angle S$ to the nearest degree, the angle that West Avenue makes with State Street.
      $\dfrac{\sin S}{53.2} = \dfrac{\sin 40°}{68.4}$; so $\angle S = 30°$
   c. Write an expression for the angle that West Avenue makes with Elm Street.
      Possible answer: 180° − (30° + 40°) = 110°. Or, if student uses the Law of Sines, check the work.

3. West Avenue makes angles of 55° with Main Street and 80° with Third Street. The distance from Main to Third along West Avenue is 40 yards. Hoa contracts to design a pocket park for the acute-angled triangular area enclosed by these streets.
   a. What is the measure of the third angle of the triangular area? **45°**
   b. Write and evaluate an expression for the distance from West Avenue to Third Street along Main Street.
      $\dfrac{40 \sin 80°}{\sin 45°} = 55.7$ **yd**

Choose the letter for the best answer.

4. Hoa wants to plant palm trees 8 feet apart along the side of the park on Third Street. Which expression gives the number of trees she will need?

   A $\dfrac{40 \sin 55°}{\sin 45°}$ C $\dfrac{5 \sin 45°}{\sin 55°}$

   (B) $\dfrac{5 \sin 55°}{\sin 45°}$ D $\dfrac{40 \sin 45°}{\sin 55°}$

5. What is the area of the new pocket park that Hoa is designing?

   (F) 913 yd^2
   G 1004 yd^2
   H 1207 yd^2
   J 1398 yd^2

41   **Holt Algebra 2**

Reading Strategy
Use a Flowchart

The procedure involved in solving a triangle depends upon the information given about the triangle.

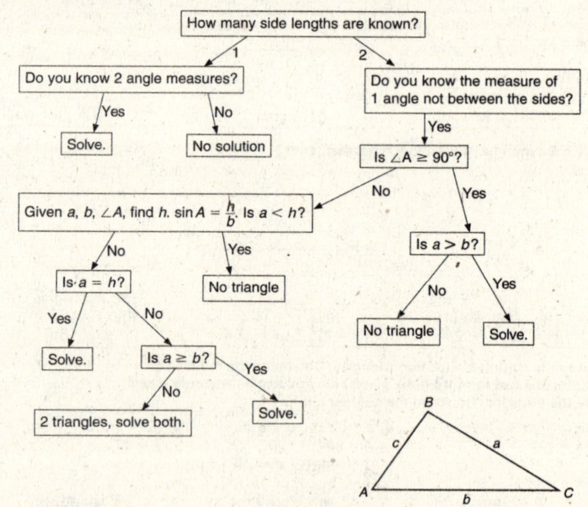

Trace a path through the flowchart for each possible triangle and state the outcome.

1. $b = 8$, $m\angle A = 55°$, $m\angle C = 45°$
   Solve.

2. $a = 13$, $b = 9$, $m\angle A = 80°$
   Solve.

3. $a = 4$, $b = 10$, $m\angle A = 95°$
   No triangle

4. $a = 8$, $b = 12$, $m\angle A = 35°$
   2 triangles

5. $a = 15$, $b = 11$, $m\angle A = 102°$
   Solve.

6. $a = 5$, $b = 7$, $m\angle A = 62°$
   No triangle

42   **Holt Algebra 2**

Holt Algebra 2

Solve each triangle. Round to the nearest tenth.

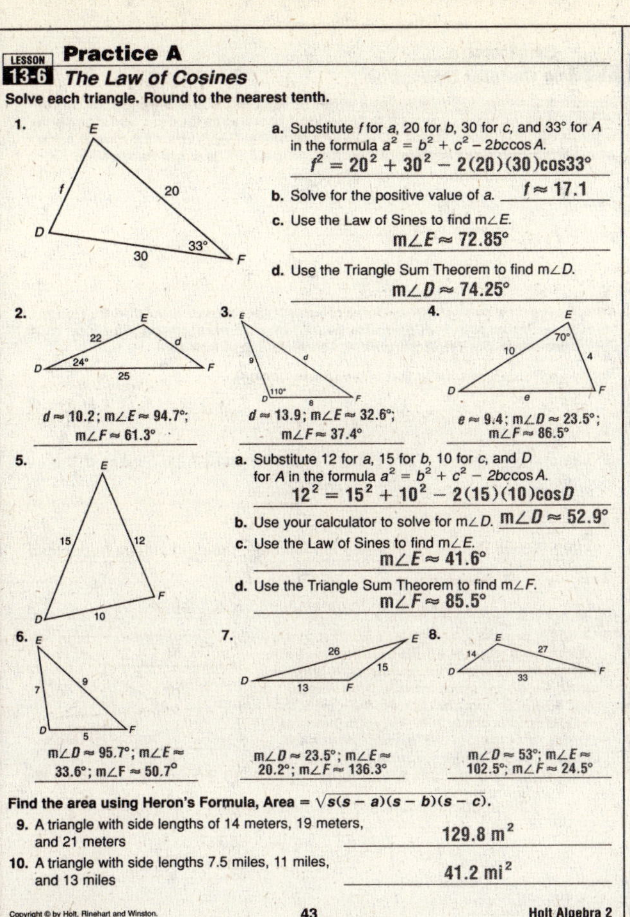

1.
a. Substitute *f* for *a*, 20 for *b*, 30 for *c*, and 33° for *A* in the formula $a^2 = b^2 + c^2 - 2bc\cos A$.
$$f^2 = 20^2 + 30^2 - 2(20)(30)\cos 33°$$
b. Solve for the positive value of *a*. $f \approx 17.1$
c. Use the Law of Sines to find m∠*E*. $m\angle E \approx 72.85°$
d. Use the Triangle Sum Theorem to find m∠*D*. $m\angle D \approx 74.25°$

2. $d \approx 10.2$; $m\angle E \approx 94.7°$; $m\angle F \approx 61.3°$

3. $d \approx 13.9$; $m\angle E \approx 32.6°$; $m\angle F \approx 37.4°$

4. $e \approx 9.4$; $m\angle D \approx 23.5°$; $m\angle F \approx 86.5°$

5.
a. Substitute 12 for *a*, 15 for *b*, 10 for *c*, and *D* for *A* in the formula $a^2 = b^2 + c^2 - 2bc\cos A$.
$$12^2 = 15^2 + 10^2 - 2(15)(10)\cos D$$
b. Use your calculator to solve for m∠*D*. $m\angle D \approx 52.9°$
c. Use the Law of Sines to find m∠*E*. $m\angle E \approx 41.6°$
d. Use the Triangle Sum Theorem to find m∠*F*. $m\angle F \approx 85.5°$

6. $m\angle D \approx 95.7°$; $m\angle E \approx 33.6°$; $m\angle F \approx 50.7°$

7. $m\angle D \approx 23.5°$; $m\angle E \approx 20.2°$; $m\angle F \approx 136.3°$

8. $m\angle D \approx 53°$; $m\angle E \approx 102.5°$; $m\angle F \approx 24.5°$

Find the area using Heron's Formula, Area $= \sqrt{s(s-a)(s-b)(s-c)}$.

9. A triangle with side lengths of 14 meters, 19 meters, and 21 meters. 129.8 m^2

10. A triangle with side lengths 7.5 miles, 11 miles, and 13 miles. 41.2 mi^2

Use the given measurements to solve each triangle. Round to the nearest tenth.

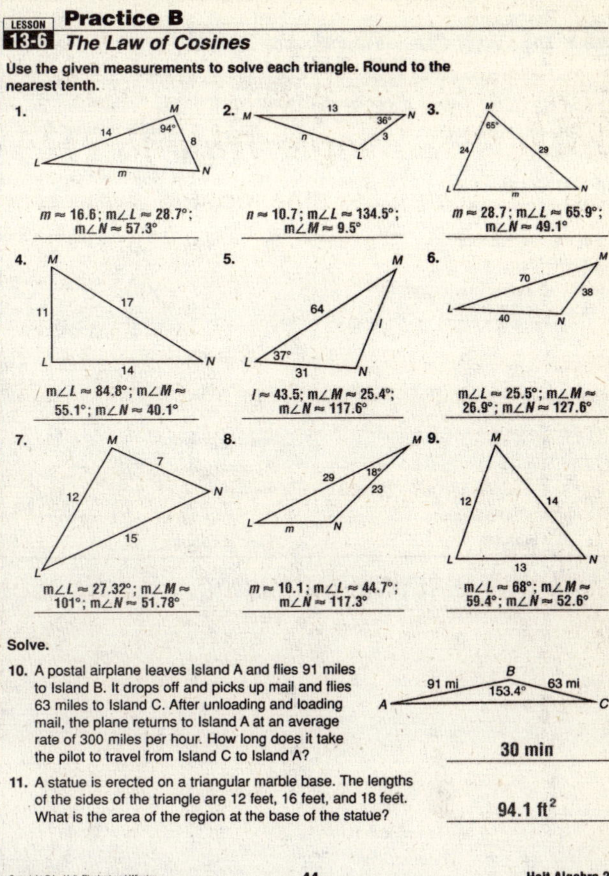

1. $m \approx 16.6$; $m\angle L \approx 28.7°$; $m\angle N \approx 57.3°$

2. $n \approx 10.7$; $m\angle L \approx 134.5°$; $m\angle M \approx 9.5°$

3. $m \approx 28.7$; $m\angle L \approx 65.9°$; $m\angle N \approx 49.1°$

4. $m\angle L \approx 84.8°$; $m\angle M \approx 55.1°$; $m\angle N \approx 40.1°$

5. $l \approx 43.5$; $m\angle M \approx 25.4°$; $m\angle N \approx 117.6°$

6. $m\angle L \approx 25.5°$; $m\angle M \approx 26.9°$; $m\angle N \approx 127.6°$

7. $m\angle L \approx 27.32°$; $m\angle M \approx 101°$; $m\angle N \approx 51.78°$

8. $m \approx 10.1$; $m\angle L \approx 44.7°$; $m\angle N \approx 117.3°$

9. $m\angle L \approx 68°$; $m\angle M \approx 59.4°$; $m\angle N \approx 52.6°$

Solve.

10. A postal airplane leaves Island A and flies 91 miles to Island B. It drops off and picks up mail and flies 63 miles to Island C. After unloading and loading mail, the plane returns to Island A at an average rate of 300 miles per hour. How long does it take the pilot to travel from Island C to Island A? 30 min

11. A statue is erected on a triangular marble base. The lengths of the sides of the triangle are 12 feet, 16 feet, and 18 feet. What is the area of the region at the base of the statue? 94.1 ft^2

Use the given measurements to solve each triangle. Round to the nearest tenth.

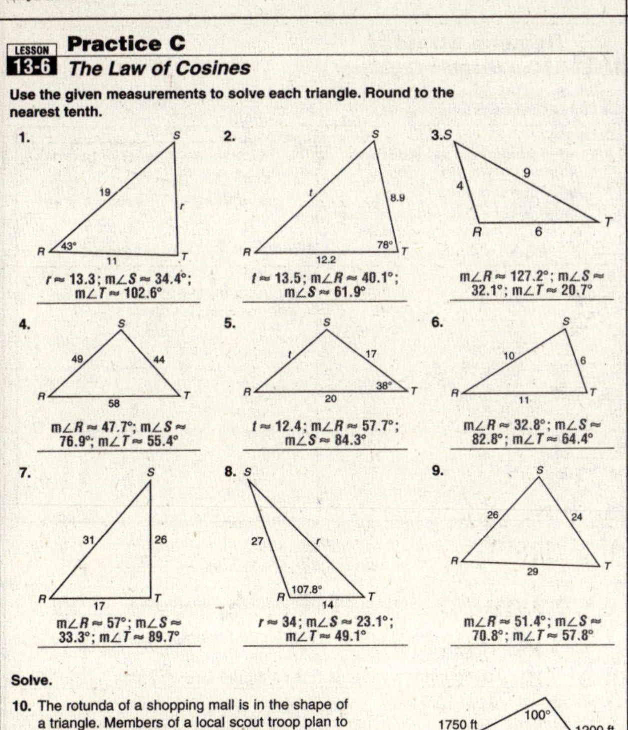

1. $r \approx 13.3$; $m\angle S \approx 34.4°$; $m\angle T \approx 102.6°$

2. $t \approx 13.5$; $m\angle R \approx 40.1°$; $m\angle S \approx 61.9°$

3. $m\angle R \approx 127.2°$; $m\angle S \approx 32.1°$; $m\angle T \approx 20.7°$

4. $m\angle R \approx 47.7°$; $m\angle S \approx 76.9°$; $m\angle T \approx 55.4°$

5. $t \approx 12.4$; $m\angle R \approx 57.7°$; $m\angle S \approx 84.3°$

6. $m\angle R \approx 32.8°$; $m\angle S \approx 82.8°$; $m\angle T \approx 64.4°$

7. $m\angle R \approx 57°$; $m\angle S \approx 33.3°$; $m\angle T \approx 89.7°$

8. $r \approx 34$; $m\angle S \approx 23.1°$; $m\angle T \approx 49.1°$

9. $m\angle R \approx 51.4°$; $m\angle S \approx 70.8°$; $m\angle T \approx 57.8°$

Solve.

10. The rotunda of a shopping mall is in the shape of a triangle. Members of a local scout troop plan to walk around the outer edge of the rotunda 5 times to increase awareness of physical fitness. If they walk at an average speed of 3 miles per hour, how long should it take them to the nearest minute? $1 \text{ h } 39 \text{ min}$

11. There is a fountain in a triangular garden in front of City Hall. The lengths of the sides of the triangle are 14 meters, 15 meters, and 21 meters. What is the area of the garden? 104.9 m^2

Use the **Law of Cosines** to solve a triangle if you know the length of two sides and the measure of the angle between them or the length of all the sides.

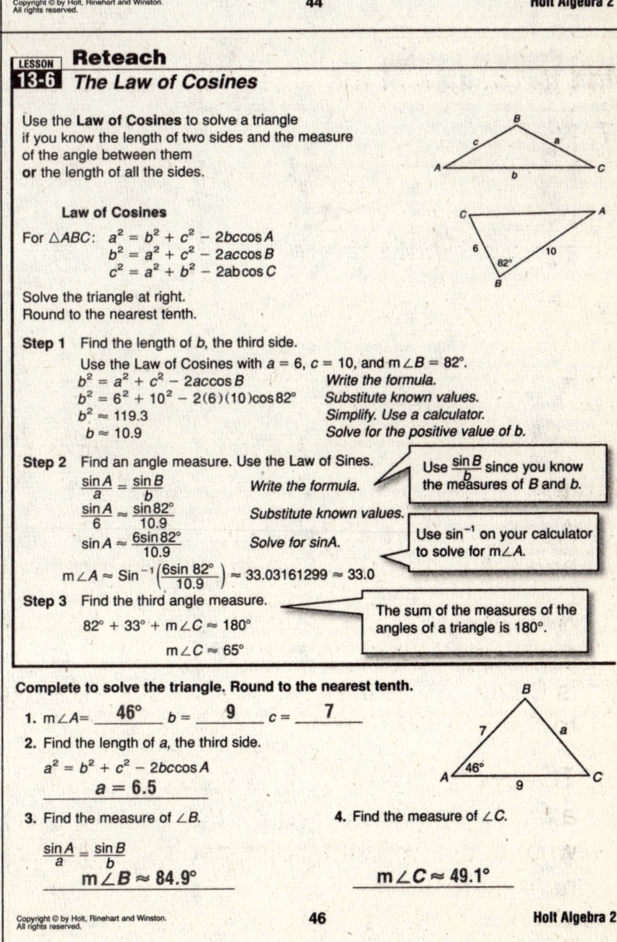

Law of Cosines

For $\triangle ABC$: $a^2 = b^2 + c^2 - 2bc\cos A$
$b^2 = a^2 + c^2 - 2ac\cos B$
$c^2 = a^2 + b^2 - 2ab\cos C$

Solve the triangle at right. Round to the nearest tenth.

Step 1 Find the length of *b*, the third side.
Use the Law of Cosines with $a = 6$, $c = 10$, and $m\angle B = 82°$.
$b^2 = a^2 + c^2 - 2ac\cos B$ *Write the formula.*
$b^2 = 6^2 + 10^2 - 2(6)(10)\cos 82°$ *Substitute known values.*
$b^2 = 119.3$ *Simplify. Use a calculator.*
$b = 10.9$ *Solve for the positive value of b.*

Step 2 Find an angle measure. Use the Law of Sines.
$\dfrac{\sin A}{a} = \dfrac{\sin B}{b}$ *Write the formula.*
$\dfrac{\sin A}{6} = \dfrac{\sin 82°}{10.9}$ *Substitute known values.*
$\sin A = \dfrac{6\sin 82°}{10.9}$ *Solve for sin A.*
$m\angle A \approx \sin^{-1}\left(\dfrac{6\sin 82°}{10.9}\right) \approx 33.03161299 \approx 33.0$

Use $\dfrac{\sin B}{b}$ since you know the measures of *B* and *b*.

Use $\sin^{-1}$ on your calculator to solve for m∠*A*.

Step 3 Find the third angle measure.
$82° + 33° + m\angle C \approx 180°$

The sum of the measures of the angles of a triangle is 180°.

$m\angle C \approx 65°$

Complete to solve the triangle. Round to the nearest tenth.

1. $m\angle A =$ 46° $b =$ 9 $c =$ 7

2. Find the length of *a*, the third side.
$a^2 = b^2 + c^2 - 2bc\cos A$
$a = 6.5$

3. Find the measure of ∠*B*.
$\dfrac{\sin A}{a} = \dfrac{\sin B}{b}$
$m\angle B \approx 84.9°$

4. Find the measure of ∠*C*.
$m\angle C \approx 49.1°$

Reteach
The Law of Cosines (continued)

Heron's Formula is used to find the area of a triangle given the lengths of its sides.

Heron's Formula

Notice that s is half the perimeter of the triangle.

For $\triangle ABC$, with $s = \frac{1}{2}(a + b + c)$,

$$\text{Area} = \sqrt{s(s - a)(s - b)(s - c)}$$

To find the area of the triangle at right, use Heron's Formula since the lengths of all the sides are known.

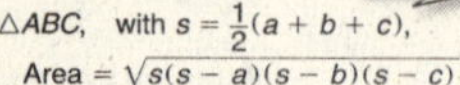

Step 1 Find the value of s.

$a = 27$, $b = 31$, and $c = 18$

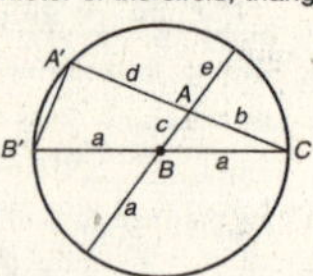

$s = \frac{1}{2}(a + b + c)$	*Write the formula.*
$s = \frac{1}{2}(27 + 31 + 18)$	*Substitute known values.*
$s = 38$	*Simplify.*

Step 2 Find the area. Use Heron's Formula. Round to the nearest foot.

$A = \sqrt{s(s - a)(s - b)(s - c)}$	*Write the formula.*
$A = \sqrt{38(38 - 27)(38 - 31)(38 - 18)}$	*Substitute. Use $s = 38$.*
$A = \sqrt{38(11)(7)(20)} = \sqrt{58{,}520}$	*Simplify.*
$A \approx 241.9090738$	*Find the square root.*
$A \approx 242 \text{ ft}^2$	*Round the answer.*

Use a calculator.

Complete to find the area. Round to the nearest whole unit.

5. $a = \underline{\quad 9 \quad}$ $b = \underline{\quad 12 \quad}$ $c = \underline{\quad 7 \quad}$

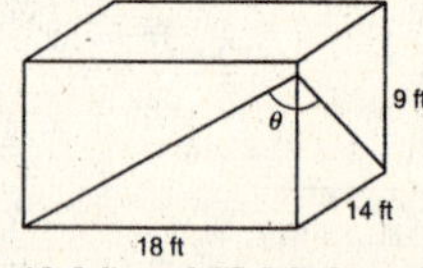

6. Find s.

$$s = \frac{1}{2}(a + b + c)$$

$$\underline{\qquad s = 14 \qquad}$$

7. Find the area.

$$A = \sqrt{s(s - a)(s - b)(s - c)}$$

$$\underline{\qquad A = 31 \text{m}^2 \qquad}$$

Holt Algebra 2

Challenge
A Circular Derivation

The Law of Cosines can be derived by a geometric argument. Consider triangle ABC below. Vertex B is placed at the center of a circle, vertex C is located on the circle, and side a is the radius of the circle. Sides AC and BC are extended so that they intersect the circle at points A' and B'. Since $B'C$ is a diameter of the circle, triangle $A'B'C$ is a right triangle since $\angle A'$ intercepts a semicircle.

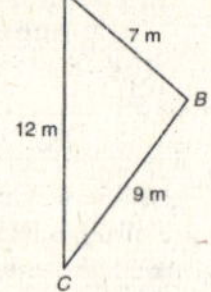

A theorem from geometry states that if two chords intersect inside a circle, then the product of the two segments of one chord is equal to the product of the two segments of the other chord. For the two chords that intersect at point A, we therefore have $bd = (a + c)e$.

Justify each step in the derivation of the Law of Cosines.

1. $bd = (a + c)(a - c)$

$\underline{\qquad c + e = a, \text{ so } e = a - c \qquad}$

2. $b(2a\cos C - b) = a^2 - c^2$

$\cos C = \dfrac{\text{adj.}}{\text{hyp.}} = \dfrac{d + b}{2a}$, so $2a\cos C = d + b$, and $2a\cos C - b = d$; the right side of the equation is the FOIL method.

3. $2ab\cos C - b^2 = a^2 - c^2$

Distributive Property simplifies the left side; the right side is unchanged.

4. $c^2 = a^2 + b^2 - 2ab\cos C$

Solve for c^2 to obtain the Law of Cosines.

5. A worker is installing an infrared security scanning device in the corner of a living room 6 inches below the ceiling as shown in the figure at right. She needs to set the scan angle on the scanning device so that it will scan from one corner of the room to the opposite corner of the room. The room measures 18 feet by 14 feet and is 9 feet high. What is the angle θ that the device should be set to scan?

Appropriate diagonals are of length 16.4 ft, 19.9 ft, and 22.8 ft; from the Law of Cosines, the angle is then $\theta = \cos^{-1}\left(\dfrac{c^2 - a^2 - b^2}{-2ab}\right)$

$$= \cos^{-1}\left(\dfrac{22.8^2 - 19.9 - 16.4^2}{-2(19.9)(16.4)}\right) \approx 77°.$$

Holt Algebra 2

Problem Solving
The Law of Cosines

Standing on a small bluff overlooking a local pond, Clay wants to calculate the width of the pond.

1. From point C, Clay walks the distances $\overline{CA}$ and $\overline{CB}$. Then he measures the angle between these line segments.

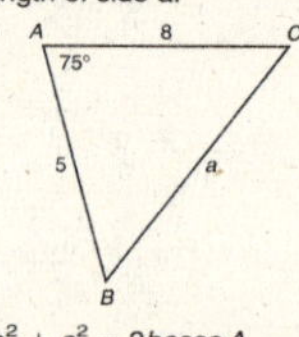

  a. Use the Law of Cosines to write an equation for the distance $\overline{AB}$.

$$c^2 = 72^2 + 60^2 - 2(72)(60)\cos 105°$$

  b. What is the distance to the nearest meter from A to B?

$$\underline{\qquad 105 \text{ m} \qquad}$$

2. From another point F, Clay measures 20 meters to D and 50 meters to E. Reece says that last summer this area dried out so much that he could walk the 49 meters from D to E.

  a. Use the Law of Cosines to write an equation for the measure of the angle between $\overline{DF}$ and $\overline{EF}$.

$$\angle F = \cos^{-1}\left(\dfrac{(50^2 + 20^2 - 49^2)}{2(50)(20)}\right)$$

  b. What is the measure of this angle?

$$\underline{\qquad 75.6° \qquad}$$

3. Reece tells Clay that when the area defined by $\triangle DEF$ dries out, it becomes covered with a grass native to this area of the country. Clay wants to know the area of this section.

  a. Use Heron's Formula to write an expression for the area.

$$A = \sqrt{59.5(9.5)(39.5)(10.5)}$$

  b. Find the area to the nearest tenth of a square meter.

$$\underline{\qquad 484.2 \text{ m}^2 \qquad}$$

A local naturalist says that a triangular area on the north sides of the pond is a turtle habitat. The lengths of the sides of this area are 15 meters, 25 meters, and 36 meters. Choose the letter for the best answer.

4. What is the area of this triangular turtle habitat?

  (A) 150.7 m²
  B 213.2 m²
  C 1012.9 m²
  D 3075 m²

5. Which expression gives the measure of the angle between the sides of the habitat that measure 15 meters and 25 meters?

  (F) $\cos^{-1}(-0.595)$
  G $\cos(-0.595)$
  H $\cos(0.595)$
  J $\cos^{-1}(0.595)$

Holt Algebra 2

Reading Strategy
Use a Graphic Organizer

A graphic organizer can be used to organize what you know about using the **Law of Cosines** to solve triangles.

Definition	Use
The **Law of Cosines** can be used to solve triangles when you are given either: • the lengths of all 3 sides, or • the lengths of 2 sides and the measure of the angle between those 2 sides. $a^2 = b^2 + c^2 - 2bc\cos A$	This law can be used to solve for any side of a triangle, a, b, or c. $a^2 = b^2 + c^2 - 2bc\cos A$ $b^2 = a^2 + c^2 - 2ac\cos B$ $c^2 = a^2 + b^2 - 2ab\cos C$
Example Find the length of side a. $a^2 = b^2 + c^2 - 2bc\cos A$ $a^2 = 8^2 + 5^2 - 2(8)(5)\cos 75° = 68.3$ $a = 8.3$	**Hint** You can use the Law of Cosines to check results from using the Law of Sines.

Answer each question.

1. Can you find the lengths of all 3 sides of a triangle if you know the measure of 2 angles? Explain.

No; there are an infinite number of possibilities for the side lengths.

2. Describe how you can check your answer when you use the Law of Cosines to solve a triangle.

You can check your answer using the Law of Sines.

3. Jenna is given the following information about triangle EFG: $e = 9$ cm, $f = 6$ cm, and $\angle G = 70°$.

  a. Write an equation she can use to find the length of side g.

$$g = \sqrt{81 + 36 - 108\cos 70°}$$

  b. Once she has found the length of side g, describe how she can find the measures of $\angle E$ and $\angle F$.

Possible answer: She could use the Law of Cosines or Law of Sines to find another angle, and then subtract the sum of the two angles from 180° to find the third angle.

Holt Algebra 2

Holt Algebra 2